I0888912

INFUSING ETHICS INTO THE DEVELOPMENT OF ENGINEERS

Exemplary Education Activities and Programs

Infusing Ethics Selection Committee

Center for Engineering Ethics and Society

NATIONAL ACADEMY OF ENGINEERING

THE NATIONAL ACADEMIES PRESS
Washington, DC
www.nap.edu

THE NATIONAL ACADEMIES PRESS 500 Fifth Street NW Washington, DC 20001

This activity was supported by grant No. 1449199 from the National Science Foundation. Any opinions, findings, conclusions, or recommendations expressed in this publication do not necessarily reflect the views of any organization or agency that provided support for the project.

International Standard Book Number-13: 978-0-309-39085-9
International Standard Book Number-10: 0-309-39085-0

Additional copies of this report are available for sale from the National Academies Press, 500 Fifth Street NW, Keck 360, Washington, DC 20001; (800) 624-6242 or (202) 334-3313; www.nap.edu.

Printed in the United States of America

Suggested citation: National Academy of Engineering. 2016. *Infusing Ethics into the Development of Engineers: Exemplary Education Activities and Programs*. Washington: National Academies Press.

The National Academies of
SCIENCES · ENGINEERING · MEDICINE

The **National Academy of Sciences** was established in 1863 by an Act of Congress, signed by President Lincoln, as a private, nongovernmental institution to advise the nation on issues related to science and technology. Members are elected by their peers for outstanding contributions to research. Dr. Ralph J. Cicerone is president.

The **National Academy of Engineering** was established in 1964 under the charter of the National Academy of Sciences to bring the practices of engineering to advising the nation. Members are elected by their peers for extraordinary contributions to engineering. Dr. C. D. Mote, Jr., is president.

The **National Academy of Medicine** (formerly the Institute of Medicine) was established in 1970 under the charter of the National Academy of Sciences to advise the nation on medical and health issues. Members are elected by their peers for distinguished contributions to medicine and health. Dr. Victor J. Dzau is president.

The three Academies work together as the **National Academies of Sciences, Engineering, and Medicine** to provide independent, objective analysis and advice to the nation and conduct other activities to solve complex problems and inform public policy decisions. The Academies also encourage education and research, recognize outstanding contributions to knowledge, and increase public understanding in matters of science, engineering, and medicine.

Learn more about the National Academies of Sciences, Engineering, and Medicine at **www.national-academies.org.**

Center for Engineering Ethics and Society Advisory Group

National Academy of Engineering

Acknowledgment of Reviewers

The summary of this publication has been reviewed in draft form by individuals chosen for their diverse perspectives and technical expertise, in accordance with procedures approved by the National Academies of Sciences, Engineering, and Medicine. The purpose of this independent review is to provide candid and critical comments to assist the NAE in making its published report as sound as possible and to ensure that the manuscript meets institutional standards for objectivity, evidence, and responsiveness to the project's charge. The review comments and draft manuscript remain confidential to protect the integrity of the deliberative process. We thank the following individuals for their review of the manuscript:

Paul Citron, Retired Vice President, Technology Policy and Academic Relations, Medtronic, Inc.
Glenn Daigger, President, One Water Solutions, LLC
Christopher Swan, Associate Professor, Department of Civil and Environmental Engineering, Tufts University
Daniel Vallero, Adjunct Professor, Pratt School of Engineering, Duke University

Although the reviewers listed above have provided many constructive comments and suggestions, they were not asked to endorse the views expressed in the summary, nor did they see the final draft of the summary before its release. The review of this publication was overseen by Proctor Reid, Director of the Program Office at the NAE. He was responsible for making certain that an independent examination of this manuscript was carried out in accordance with institutional procedures and that all review comments were carefully considered. Responsibility for the final content of this publication rests entirely with the authors and NAE.

Preface

Ethical practice in engineering is critical for ensuring public trust in the field and in its practitioners, especially as engineers increasingly tackle international and socially complex problems that combine technical and ethical challenges. This report aims to raise awareness of the variety of exceptional programs and strategies for improving engineers' understanding of ethical and social issues and provides a resource for those who seek to improve the ethical development of engineers at their own institutions.

Ethics is of crucial importance to the engineering profession, as evidenced both in the codes of ethics published by numerous engineering professional societies and in the requirements for accredited engineering programs maintained by the US Accreditation Board for Engineering and Technology (ABET). According to the ABET criteria, students in accredited programs must demonstrate an understanding of ethics and take it into account when designing a system, component, or process. The ABET requirement applies to both undergraduate and graduate programs in engineering and engineering technology and has spurred schools to provide engineering ethics education for their students in a variety of ways.

A number of these engineering ethics education activities were reviewed for this project, with the goal of selecting and widely disseminating those that may serve as exemplars for broader adoption and adaptation. They were gathered by the advisory group for the National Academy of Engineering (NAE) Center for Engineering Ethics and Society (CEES), which invited faculty and administrators at US universities and colleges to submit activities that prepare students for ethical practices, research, or leadership in engineering. Eligible activities were those at the associate's, bachelor's, or master's level for engineering or engineering technology. Additional information about and materials from the exemplars in this report will be included in the NAE's Online Ethics Center for Engineering and Science (OEC) collection (onlineethics.org).

Funded by the National Science Foundation, this effort builds on two other NAE reports on engineering ethics education, *Practical Guidance on Science and Engineering Ethics Education for Instructors and Administrators* (NAE 2013) and *Ethics Education and Scientific and Engineering Research: What's Been Learned? What Should be Done?* (NAE 2009). This project also aligns with NAE efforts to improve engineering education,[1] prepare engineers for the future,[2] and educate engineers to address far-reaching and fundamental engineering challenges.[3]

A specially appointed NAE selection committee reviewed the submissions and identified programs that serve as examples for those who wish to prepare engineers to think critically about the ethics of their profession. The following seven members served on the Infusing Ethics Selection Committee:

Stephanie J. Bird, ethics consultant and coeditor of *Science and Engineering Ethics*
Andrene Bresnan, director, Ethics and Business Conduct, The Boeing Company
Gerald E. Galloway, Jr., Glenn L. Martin Institute Professor of Engineering, University of Maryland, College Park
Joseph R. Herkert, visiting scholar, Genetic Engineering and Society Center, North Carolina State University
Sharon D. Kenny, civil engineer and project management professional
Indira Nair, professor and vice provost emerita, Carnegie Mellon University
Chris Schairbaum, director, Innovation and Development, Texas Instruments, Inc.

The committee members were impressed by the variety and quality of the submissions and excited to see the creative approaches to infusing ethics into the development of engineers.

The 25 *NAE Exemplars in Engineering Ethics Education* described in this report serve as a resource for institutions and educators to strengthen and expand their ethics programs and thus improve the capabilities of practicing and future engineers. The NAE is very pleased to acknowledge these efforts and encourages engineering educators and practitioners to consider and incorporate these strategies.

Dr. C. D. Mote, Jr.	Dr. Gerald E. Galloway
President	Chair
National Academy of Engineering	CEES Advisory Group

---

[1] Frontiers of Engineering Education, www.naefoee.org/
[2] *The Engineer of 2020* (2004); *Educating the Engineer of 2020* (2005).
[3] Grand Challenge Scholars Program, www.engineeringchallenges.org/GrandChallengeScholarsProgram.aspx

Table of Contents

Summary

Project Summary

This publication presents 25 activities and programs that are exemplary in their approach to infusing ethics into the development of engineering students. It is intended to serve as a resource for institutions of higher education seeking to enhance their efforts in this area. The National Academy of Engineering's (NAE) Center for Engineering Ethics and Society Advisory Group and Infusing Ethics Selection Committee acknowledge the leadership of the National Science Foundation in funding both this project[1] and a follow-up workshop[2], and for funding much of the research that led to the ethics education activities described in many of these programs.

A call for submissions, sent to deans, chairs, and faculty in engineering, engineering technology, the social sciences, and the humanities, yielded 44 submissions.[3] Submissions had to meet two criteria: the activity should connect ethics to technical engineering content and should include assessment, quantitative or qualitative, of whether its education goals have been or are being met. Within these broad parameters, submissions ranged from short activities inserted in engineering courses to multiyear programs required of all students. The list of all submissions (appendix A) shows the variety of ideas and approaches for engineering ethics education.

In assessing the submissions the members of the selection committee looked for the following characteristics:

- Provides an interactive format that encourages active learning
- Occurs across multiple years of a student's education
- Includes an institutional faculty reward structure that supports ethics training
- Connects students' ethics learning to engineering practice
- Promotes improved ethical decision-making and problem-solving skills
- Addresses <u>macroethics</u> (the broader ethical and social issues that call for the collective response of the engineering profession and societal decisions about technology), <u>microethics</u> (ethical issues involving the interactions and individual actions of engineers in research and practice), or <u>both</u>[4]
- Incorporates innovative or creative educational methods
- Has a demonstrated widespread or lasting impact on students
- Can be scaled up or easily replicated at other institutions

The 25 selected programs were picked because they clearly exhibit at least one and typically several of these features. The committee also considered the variety of educational approaches and topics covered. The exemplars presented in the following pages encompass a range of program types—undergraduate and graduate courses, multiyear programs, extracurricular experiences—and institutions to illustrate the diversity of effective approaches to infusing ethics into engineering education.

The committee recognizes that incorporating ethics education activities into the very full engineering curriculum can be more challenging than knowing what activity or program to institute, and so submitters were asked to comment on their experiences and to offer suggestions for overcoming these challenges. The comments and suggestions are summarized below.

This publication is intended as a resource for engineering and engineering technology faculty and administrators interested in approaches for creating, strengthening, or expanding their programs for infusing ethics into the development of engineers. Further information that will enable faculty and administrators to replicate and adapt the programs will be made available in spring 2016 on the Online Ethics Center for Engineering and Science (onlineethics.org).

Organization of This Publication

The descriptions begin on page 3, organized in the following broad categories: graduate course, undergraduate course, multiyear program (in which students participate at multiple times during their college education), and other (workshop, extracurricular program, and faculty development program).

The write-up for each program/activity names the faculty and institution(s) involved, describes exemplary features of the program/activity, and provides assessment information and additional resources. Appendix A provides a full list of submitted programs and activities that are working to infuse ethics into the development of engineers, and a map in appendix B shows

---

[1] This material is based on work supported by the National Science Foundation under Grant No. 1449199. Any opinions, findings, and conclusions or recommendations expressed in this material are those of the authors and do not necessarily reflect the views of the National Science Foundation.

[2] This project is supported by the National Science Foundation under Grant No. 1550637. More information is available at http://nsf.gov/awardsearch/showAward?AWD_ID=1550637&HistoricalAwards=false

[3] Two submissions did not qualify for evaluation.

[4] Herkert J. 2001. Future directions in engineering ethics research. Science and Engineering Ethics 7(3):403–414.

the geographic distribution of the selected programs, revealing that there are faculty and institutions that can serve as local collaborators for most regions of the United States.

Challenges and Suggestions for Infusing Ethics

The 46 individuals who submitted ethics programs for consideration were asked to provide their comments and ideas to help others who would like to improve or create ethics training at their own institutions. Nine comments were submitted anonymously, eight from faculty members and one from an administrator.

Of the nine respondents, seven said they have faced challenges in implementing ethics activities. While this small sample does not represent a full picture of the range of challenges experienced by faculty throughout the United States, the respondents expressed sentiments that will be familiar to many educators.[5] Most noted

- a lack of interest among students (student challenges),
- resistance from faculty (faculty challenges), or
- a lack of consensus on important topics and methods for incorporating ethics in an already overstuffed curriculum (topical and pedagogical challenges).

The respondents suggested ways to overcome these challenges and offered useful advice for other faculty and administrators.

To address **student challenges**, respondents suggested that faculty use real-life and relatable examples in their classes, make sure ethics activities are interactive, and enlist working engineers as well as other engineering faculty to participate whenever possible. One faculty member noted that students were more eager to give attention to issues that their peers identified as important, and reported success with having a student develop new cases and an interactive lesson for classmates.

Faculty challenges involved resistance to colleagues' efforts to infuse ethics in the curriculum and concerns about being unprepared to teach ethics. One respondent urged administrators to ensure that educators receive adequate training and ongoing support, beginning with those most interested, and to cultivate buy-in from diverse constituencies, including students and faculty from other departments. It was suggested that initial activities be "low intensity" until a critical mass of interest could be developed, and that it would be helpful for important or recognized individuals, such as the school president, attend these early events. Faculty members were encouraged to seek partners with differing expertise for consultation, coteaching, or course design. And individuals seeking to influence the culture of their department and thereby address the challenge of resistant faculty colleagues were advised to draw on resources that others have already created and to seek out mentoring and support from experts at other institutions.

While engaging other faculty and a larger group of constituencies may help to provide support and expertise to educators who feel underprepared, it may not entirely address **topical and pedagogical challenges**. For these, suggestions focused on consensus building through conversations, workshops, and regular meetings. One respondent reported that a semiregular faculty ethics happy hour was very helpful in allowing faculty to connect and work out methods for bringing ethics topics into the classroom.

Follow-on Workshop

An NAE workshop in the second half of 2016 will invite engineering and ethics educators to address obstacles, identify solutions, and develop institutional plans for effectively incorporating ethics in engineering education, culture, and curriculum. Funded by the National Science Foundation, the workshop will provide information, guidance, and opportunities for facilitated discussion. Details on applying to attend the workshop will be available in late spring 2016 at the Center for Engineering Ethics and Society web page (www.nae.edu/Activities/Projects/CEES.aspx). Video and a published summary of the workshop will be made publicly available.

---

[5] Walczak, K., Finelli, C., Holsapple, M., Sutkus, J., Harding, T., & Carpenter, D. (2010). Institutional Obstacles to Integrating Ethics into the Curriculum and Strategies for Overcoming Them. In *ASEE Annual Conference and Exposition, Conference Proceedings*; Sunderland, M. E. (2013). Using Student Engagement to Relocate Ethics to the Core of the Engineering Curriculum. *Science and Engineering Ethics*, 1–18; Newberry, B. (2004). The dilemma of ethics in engineering education. *Science and Engineering Ethics*, *10*(2), 343–351.

Responsibility of Engineering: Codes and Professionalism (3-hour university course)

KANSAS STATE UNIVERSITY

College of Engineering

Institution: Kansas State University

Faculty/contributors: Steve Starrett[1]

Exemplary features: Leverages student work experiences; interaction with practicing engineers

Why it's exemplary: Students in this 3-hour graduate course gain valuable insight into engineering ethics through numerous lectures and assignments. The assignments call for conducting a formal interview of engineers about ethical challenges they have faced, creating a two-hour engineering ethics workshop, and authoring a personal engineering code of ethics. Students who are practicing engineers, as well as beginning full-time graduate students, have found the formal interview very helpful. The students also have to create a workshop for their peers, with roundtable activities. For the engineering codes of ethics, which students create at the end of the semester, students gain ideas from professional societies' codes of ethics and then write a code of ethics that encompasses aspects that are vitally important to how they desire to live as an engineer.

Program description: This activity is a complete, graduate-level, 3-hour university course on engineering ethics. I created it about 15 years ago and have been teaching it continuously ever since. The focus is on the engineer as an individual designer, consultant, inspector, contractor, vendor, and/or government employee. Most of the multiple hundreds of students who take this course are practicing engineers with a few years of experience; some have more than 10 years' experience. They mostly have backgrounds in civil engineering, electrical engineering, computer engineering, industrial engineering, and engineering management. Most students have had some formal exposure to engineering ethics—in a seminar-type course, as part of a day's discussion in a technical design course, during a few sessions in a senior project course, or through a lunch discussion hosted by a local chapter of engineers.

This CE 703 course is an intense study exclusively focused on the ethical responsibilities of being an engineer. The practicing engineers who take it have encountered ethical dilemmas and can highly relate to the subject matter of the course. Their practical experience maximizes the learning potential of the course. One of the assignments is for students to analyze an ethical dilemma they faced or were aware of. The complexity of the situations described is vast and difficult with no easy way to solve the ethical dilemmas. Some example situations are a politically appointed leader repeatedly overrules a county engineer over the closure of a highly dangerous bridge (i.e., most wooden piers completely rotted through); a company placing so much emphasis on its stellar safety record that accidents are not reported so hazards in a work environment go unaddressed (i.e., no one wants to be "that person" who breaks the record of accident-free days; a bribe is demanded of an engineer in return for a good inspection while working in a foreign environment (i.e., the project may die without a good inspection); engineers are asked to design a public facility with inadequate funds (i.e., facility can't be safely built with the resources available); engineers are pressured to adjust a technical report to decrease anticipated negative effects on the environment; and engineers are pressured to keep quiet after determining that poor record keeping by governmental utilities caused a client's underpayment for utility services by millions of dollars. The undergraduates who take this course have limited practical experience to build on so this course presents a lot of new aspects that they have not thought of before. These examples illustrate the need for engineering ethics education. The students who faced these dilemmas stated how incredibly valuable CE 703 Responsibility of Engineering: Codes & Professionalism was to them.

Key goal: Increase abilities to make solid ethical decisions. Engineers need to approach ethical dilemmas using their strengths, similar to how they approach technical problems. Sometimes engineers give in to unethical solutions proposed by others because they think it's a nontechnical problem so others are better prepared to handle the issue. This course provides students new knowledge and the awareness of different engineering ethical concepts and approaches (i.e., character based, principle based, consequence based) to be used when facing ethical dilemmas. The NSPE and other codes of ethics are studied in detail. Students gain considerable confidence in their abilities to make highly ethical decisions by the end of the course. I don't expect this course alone to change an "unethical" engineer into one with high standards. I also don't take credit for the high ethical standards most engineers have that take this course. The students are interested in engineering ethics and enroll in a graduate-level engineering ethics course, indicating that they have a high level of appreciation for this topic already. I also understand that students who complete this course may make unethical decisions: they may succumb to the intense pressure of lack of time, peer pressure to make a bad decision, a supervisor requiring an unethical decision, the pursuit of profit, an unethical team culture, or the pursuit of being famous. I do expect students who complete CE 703 Responsibility of Engineering: Codes & Professionalism to be better able to understand which alternatives are ethically acceptable and which are not, champion ethical solutions

[1] Faculty/contributors are affiliated with the host/primary institution unless otherwise indicated.

when part of a team faced with an ethical dilemma, and generally conduct themselves as a professional with high standards.

Assessment information: Formal and informal feedback from students has clearly indicated how much students have valued CE 703 Responsibility of Engineering: Codes & Professionalism. The following feedback has been documented in assessments: "Great teacher, great course." "The instructor for this course is excellent. The primary purpose for taking this course was twofold; I had this instructor in a previous leadership and diversity class, which I found to be an excellent leadership experience, and secondly I wanted to learn more about improving my awareness of ethics and leadership skills. This course was one of the best learning experiences that I have had since returning to college to obtain my graduate degree. I wish that I had been taught this subject material years ago—it would have made my transition to management a lot smoother." "The class was really useful for me, it is a subject that engineers do not pay any attention to. Everyone thinks that the classes that matter are the ones that are heavy on the mathematical side and all, but we need more classes like this one so we can become a better engineer and a better person."

Some informal (email or verbal communication) feedback has been: "This is the most important course I have ever taken." "Every engineering student should be required to take this engineering ethics course." "This course has changed my life, I am much better prepared to make solid ethical decisions when faced with dilemmas." "After completing this course I'm much better prepared to mentor the engineers that report to me." The IDEA Center course evaluation process was used to formally assess this course. Four sections when the full IDEA diagnostic form was used are reported here (2012–2015). The IDEA Short Form was also periodically used, so the assessment results were not combined between the two different assessment methods. Students were asked to rate the question of "excellent course" on a scale of 1 to 5; a rating of 4 or 5 indicates the students rated the course as excellent. The average rating over the last four years was 4.2. Progress on the essential learning objective of "developing a clearer understanding of, and commitment to, personal values" was also assessed; students on average rated their progress on this objective as substantial or exceptional. This high level of progress ranked the course on this specific learning objective nationally in the top 10% of all engineering courses that used the IDEA Center assessment tools (a converted average score of 65; a score higher than 62 places the result in the highest comparison category). I have taught university courses for 21 years. I am not aware of a single other engineering ethics 3-hour course taught online at the graduate level for practicing engineers. The continual, positive feedback I get from current and past CE 703 students communicates the important knowledge that students receive from taking the course.

Using Student-Authored Case Studies to Teach Bioengineering Ethics

Institution: University of Pittsburgh

Faculty/contributors: Rosa Lynn Pinkus

Exemplary features: Reproducible approach; teaches how to frame an ethics issue; involvement of class alumni

PITT | SWANSON ENGINEERING
BIOENGINEERING

Why it's exemplary: The use of case-based reasoning to teach professional ethics is currently recognized as a "best practice," but there is still much to be understood regarding how professionals learn ethics using case-based reasoning. It is unclear what types of cases should be used in teaching and often difficult to assess what students actually learn when reasoning with cases. The exemplar activity described here requires graduate students in a 14-week bioengineering ethics course to author their own case study, based on their research, and analyze it using methods taught in the course. Based on solid research from two consecutive NSF grants, this activity is exemplary as it recognizes the "ill-defined" nature of ethics and teaches students how to "frame" an ethics issue. This skill is needed to grapple with everyday practical ethics issues which are not neatly packaged into teaching cases. These skills have an "afterlife" when the course concludes.

Program description: The participants in the activity are graduate and undergraduate students as well as postdoctoral students and select bioengineering faculty. The course has been taught for 19 years. Former students who are employed locally as assistant professors, or employees of device companies, have the opportunity to return to the class to teach their case study and to reflect on "ethics in practice." They serve as "role models" for the younger students and in turn deepen their appreciation of the importance of ethics in their daily work. Two former students, now faculty, currently teach an undergraduate ethics course and a professional MS course. The educational goals of the 14-week required graduate bioengineering ethics are designed to supplement students' traditional bioengineering education in three specific ways:

- To teach the conceptual tools needed to identify, articulate, and resolve ethical dilemmas inherent in the practical, professional work that they take part in daily.
- To enable students to recognize that engineering and medicine each have unique demands related to professional practice, but each practice also reflects the general societal values in which the practice occurs.
- To provide a learning environment where practical ethics can best be understood. This includes group discussion of relevant ethics cases, field trips so students can observe professional ethics in action, and peer-reviewed, in-class group projects.

During the first few weeks of the course, students engage in small-group, case-based discussion. They learn to identify the ethical dilemma(s) and the morally relevant facts and concepts in their assigned case, discuss alternatives resolutions, and justify a resolution. Assigned readings and class discussions begin to introduce paradigm case studies. Key concepts such as "ethics," "professional ethics," "personal ethics," "role morality," ethical principles, and theories are defined. This "steep start-up time" provides a common language for students and sets the tone for the class. Students, engaged in discussion, begin to ask how the cases can be resolved! The next portion of the course also employs small-group discussion while introducing students to a variety of methods of moral reasoning. These methods are used as students discuss complex, technical bioengineering cases and classic paradigm cases in bioethics, tailored to their professional domains. This begins a more formal practice in using various methods. Select students from previous years coteach during this time and present their case studies. Students in the class are introduced to multiple knowledge domains in bioengineering as well as a range of ethical dilemmas and methods of moral reasoning. (Issues covered typically match those identified in the 2009 FDA Guidelines for Responsible Conduct of Research.) The final weeks of the class are devoted to presentations of their student-authored cases. The capstone assignment is writing and analyzing a case study based on the student's research area and presenting it in class. Students are instructed to work in teams of two to four. They are coached on various ways to combine their knowledge domains, so that the final case represents a blend of real-life experiences, professional knowledge, and ability to use a method of moral reasoning. The case may be written using moral imagination. If a student is working on a device that is years from going to market, s/he can "fast-forward" it and foresee possible dilemmas that might occur after it is in use. If the case is based on an actual dilemma, such as an issue with one's mentor or a problem in the lab, the facts are changed so that confidentiality is preserved. Analysis of the case must use moral problem-solving methods discussed in class. The goal of this assignment is to provide students with the experience of applying one or more methods of moral analysis taught in the class to their own work. It also encourages them to specify facts and concepts of the case and frame the dilemma so as to view their research through a professional ethics lens. The in-class case presentation provides students with the opportunity to engage fellow students in a discussion of their work, encourage them to identify the ethics issues, and suggest resolutions to the case. A peer-review system is used during the presentation to provide each student/group with constructive comments and to impress upon students that there is an objective way to evaluate student-authored case study. Christopher Schunn's SWoRD program, Scaffolded Writing and Rewriting in the Discipline (https://sites.google.com/site/swordlrdc/directory, accessed January 29, 2015), a web-based instructional system that supports reciprocal student authoring and student peer review, has been adapted to the task of writing, reviewing, and rewriting ethics problem analyses. The adaptation provides peer reviewers with written guidance for reviewing a peer's problem analysis, focusing on the use of five measures of moral reasoning, discussed in the assessment section of this submission. In addition to in-class instruction, students are required to attend two "out-of-class" assignments. This may include observing at the University of Pittsburgh Medical Center's (UPMC's) or UPMC Children's Hospital Ethics Committee Meeting; attending University of Pittsburgh's Center for Bioethics and Health Law's Bioethics Grand Rounds; or taking advantage of other offerings in the ethics programs at the University of Pittsburgh or Carnegie Mellon University. An option also exists for students to attend a US Food and Drug Administration (FDA) Panel Meeting in Washington, DC and observe, firsthand, the process by which a device is evaluated for market approval. The activity is funded by the Bioengineering Department at the University of Pittsburgh. Logistics are coordinated by the students and the TA of the class. Former Pitt students who work at the FDA host students and informally discuss workings of the FDA and answer questions about the panel meeting proceedings. The activities described above contribute to the creation of a "moral community" in the class, where difficult issues can be discussed openly. This explicit and tacit overall goal of the class is perhaps the most important aspect of this exemplary activity.

Assessment information: A student's final grade is based on

- Attending and participating in class (15%).
- Completing one critique/analysis of an out-of-class activity (10%). Students are instructed to observe the activity and note what the ethics issues were, if they were resolved, and how.
- Analyzing at midterm one complex technical case (10%)—the Teletronics Pacemaker Case Study (Pinkus and Bates, unpublished casebook, University of Pittsburgh, 2005).
- For the capstone assignment, writing and analyzing a case study based on the student's research area (40%) and presenting the case in class (25%).

An innovative assessment grid that includes 5 higher methods of moral reasoning is used to grade student's final papers: (1) Employs professional/technical knowledge to frame the issue. (2) Views the problem from multiple perspectives. (3) Flexibly moves among multiple perspectives. (4) Identifies analogous cases and articulates ways the cases were analogous. And (5) Employs a method of moral reasoning in conducting the analysis. Developed during work on the NSF grant referred to earlier, these are not stand-alone criteria. Taken together, they allow for various ways that students will frame, analyze, and resolve an ethics dilemma. These are also the criteria that are included in the students' peer review forms, so students are aware of how they are graded. The fifth criterion has been objectified to score whether or not a student labels, defines, and applies a concept used in the course. The concepts are specific (e.g., "risk assessment," "cost-benefit analysis") and general (e.g., "case-based reasoning," "utilitarianism"). An ethics concept is said to be "labeled" if the term for the concept is present; "defined" if a dictionary-like definition of the concept is present; and "applied" if the concept is brought to bear appropriately using facts of the particular case. Each of these could be done correctly or incorrectly. All concepts used in the course are listed on the grid and a grader can review the paper and note which individual concepts students use. This attends to the ill-defined nature of applied ethics and allows for assessment of individual framing of student-authored cases. Beyond the formal grading is the "deliverable" of the final student-authored cases. In addition to being taught in the graduate course described here, these have been used in the undergraduate ethics courses, the professional MS course, and the MD/PhD 4-week ethics workshop at the University of Pittsburgh. They are uniquely creative and speak to frontline issues that graduate students face. The fact that the students generously donate their time to teach in the ethics courses at the university, with their case as the focus of the presentation, attests to the fact they value the activity and reflect on the content. The fact that two graduates of the bioengineering program, one an assistant professor and another an employee of a device testing company, now teach ethics courses in the department also speaks to the impact of both the course and the student-authored case exercise.

Additional resources:

Goldin IM, Ashley KD, Pinkus RL. 2006. Teaching case analysis through framing: Prospects for an ITS in an ill-defined domain. In: Workshop on Intelligent Tutoring Systems for Ill-Defined Domains, 8th International Conference on Intelligent Tutoring Systems. Jhongli, Taiwan. www.cs.cmu.edu/~hypoform/its-workshop/papers/ITS06_illdefinedworkshop_GoldinEtAl.pdf

Goldin I, Pinkus RL, Ashley KD. 2015. Validity and reliability of an instrument for assessing case analysis in bioengineering ethics education. Science and Engineering Ethics. doi: 10.1007/s11948-0159644-2; http://link.springer.com/article/10.1007/s11948-015-9644-2

Pinkus RL, Gloeckner C, Fortunato A. 2015. The role of professional knowledge in case-based reasoning. Science and Engineering Ethics. doi: 10.1007/s11948-015-9645-1; http://link.springer.com/article/10.1007/s11948-015-9645-1

Learning to Listen: A Tool for Morally Engaged Engineering Practice

VirginiaTech
Invent the Future®

Institution: Virginia Polytechnic Institute and State University (Virginia Tech)

Faculty/contributors: Yanna Lambrinidou and Marc Edwards, Virginia Tech; Erin Heaney and Rebecca Newberry, Clean Air Coalition of Western New York

Exemplary features: Addresses the ethical responsibilities of engineers to engage with the public and other stakeholders

Why it's exemplary: "Learning to Listen" (L2L) teaches engineers the method of ethnographic listening to diverse publics who are affected by engineering interventions but whose voices are often ignored. It cautions that failure to consider such voices can leave engineers vulnerable to incomplete understanding of complex issues, self-interest, and institutional pressures, contributing to suboptimal professional decisions, unethical conduct, and even public harm. Based on the premise that morality is not a fixed theoretical body of knowledge that exists apart from day-to-day living and professional practice, L2L challenges the notion that comprehension of moral codes, theories, and principles alone equips engineers to determine what constitutes "ethical" professional conduct in different contexts. The training is unique in fostering ethical decision making not as abstract determinations of "right" and "wrong," but as direct engagement with local experiences, knowledges, and values, and careful assessment of what in each context constitutes appropriate use of professional power and technical expertise.

Program description: L2L is one of four thematic units in Virginia Tech's Civil and Environmental Engineering graduate-level course Engineering Ethics and the Public (CEE 5804). Emphasizing that morally sound engineering research, practice, and stewardship necessitate acute awareness of the perspectives of those affected by engineering interventions, it cross-cuts the course's three other units: Responsible Conduct of Research, Responsible Conduct of Practice, and Witnessing Wrongdoing and the Obligation to Prevent Harm. Yanna Lambrinidou, a medical ethnographer, and Marc Edwards, an environmental engineer, cofounded CEE 5804 in 2010 to highlight critical ethical lessons from a multiyear effort to understand and redress the still-unfolding effects of Washington, DC's historic lead-in-water contamination of 2001–2004. At the center of this case are residents who first suspected a serious problem with hazardous levels of lead in their drinking water and several groups of government agency engineers and scientists who helped conceal the contamination and ultimately covered up the extensive public health harm it caused. The DC story lends itself to teaching engineering ethics through real-world events and through the voices of local residents and other stakeholders who were impacted. Our course, which is offered every fall, was funded in 2011 by NSF's Ethics Education for Science and Engineering (EESE) program. Today the over 2 million practicing engineers in the United States routinely make complex and critical decisions with significant implications for the public's health, safety, and welfare in a relational vacuum, where affected publics are rarely seen and almost never heard. Yet the experiences, knowledges, and values of these publics often provide crucial insight and sometimes correction with respect to engineers' areas of technical expertise and moral responsibility. The goal of L2L is to teach that morality is not a fixed theoretical body of knowledge that exists apart from day-to-day living and professional practice. Therefore, it ought not be considered a neutral "compass" that engineers can use to determine relevant moral questions without an understanding of what is locally at stake in each case. Rather, to promote the public good in morally sound and socially just ways, engineers must complement their knowledge of moral codes, theories, and principles with empirically derived understandings of the experiences, knowledges, and values of the publics affected by their work. Similarly, they must know that institutionally sanctioned claims and histories may exclude important facts. With an expanded awareness about the complexities of a case—especially in relation to potential uncertainties and disagreements about the science, costs, practical benefits, and acceptable risks as well as power differentials among stakeholders—engineers can be in a better position to identify moral dilemmas and make thoughtful decisions about morally sound responses to them.

L2L combines (1) readings/lectures about local knowledge and the role diverse publics have played in successfully complementing, advancing, and challenging dominant paradigms of engineering/scientific thought and practice, with (2) semester-long, hands-on training in the critically important first steps of gathering the often-confusing and sometimes-concealed facts of real-world controversies involving engineering interventions. Offering in-depth ethnographic listening as a tool for empirically based understandings of the moral dimensions of a case, the training consists of three exercises. The first two prepare students for the third, the term project (assignment link provided below):

(2a) Anatomy of in-depth listening: Students write about four of their own experiences with in-depth listening: two as speakers, two as listeners, two positive, and two negative. They describe behaviors, observations, and feelings they remember, concluding with a reflection on what "good" and "bad" listening look and feel like and what effects they can have on one's capacity to express oneself or relate to others. Responses are compiled for everyone's review.

(2b) Practice of in-depth listening: Students conduct one face-to-face interview with someone they know well. They focus on understanding views that their interviewee holds but that they, themselves, find objectionable. The goal is to gain clarity on those views and the reasons behind them, while refraining from interpretation and judgment. Students are advised to ask all questions necessary to see the subject in question from their interviewee's perspective. They are reminded that their task is to understand, not necessarily to agree. Written reports provide reflection on what students learned and how they performed as interviewers. The latter assessment includes interviewee feedback as well.

(2c) In-depth listening in engineering and science: Students conduct a sustained investigation into an unfolding engineering controversy, which culminates in one in-depth ethnographic interview of an affected stakeholder (e.g., parent, grassroots community organization representative, scientist advocate) whose voice is underrepresented or misrepresented in official depictions of the case.

Final reports consist of a detailed description of the case; a discussion of key moral transgressions as identified by interviewees; "lessons learned" that changed students' original understanding of the case; reflections on the conduct of engineers/scientists in the case; and thoughts on actions the students themselves would want to have taken if they were involved. Usually each student selects a topic. In 2012, however, we partnered with the grassroots environmental health and justice organization Clean Air Coalition of Western New York (CACWNY), which at the time was a key stakeholder in an unfolding engineering controversy (syllabus and paper link provided below). Students collectively conducted extensive background research on the case and were paired up individually with local stakeholders for ethnographic interviews. This was an especially powerful experience for many because it amplified their research and personal connection to the case (student blog link provided below). Two students subsequently joined Lambrinidou on a field trip to CACWNY. The experience reinforced takeaway messages from the class, which they highlighted in a talk to CEE 5804 the following fall (video excerpt link provided). We consider our partnership with CACWNY a model and readopted it in fall 2015 with a community in Flint, MI; a colleague will adopt it in 2016 with a community in Seattle for a new engineering ethics undergraduate class at Seattle University.

Assessment information: Our ultimate goal is to facilitate a change in how students see themselves and their professional responsibility in relation to the safety, health, and welfare of the public. Although we lack long-term data on whether our activity (and course more broadly) is meeting this goal, we have some evidence that, at least in the short-term, it helps shift students' thinking in fundamental ways. We draw on two assessment questionnaires. The first, administered at the end of fall 2012, solicited student views about the main components of the course (e.g., lectures, readings). Student comments on the L2L unit revealed the following emerging themes: (a) 12 of 15 students noted that their exposure to real-world unfolding cases and the perspectives of marginalized stakeholders rendered engineering ethics "real," "meaningful," and "personal" because it gave "a face" to the ideas, concepts, and principles taught in class, making them more understandable and memorable, and inspiring self-reflection; (b) 9 of 15 students noted that their newly acquired ability to investigate a controversy ethnographically empowered them to uncover important dimensions of the case that were absent from official reports, and "brought the case home" on a deeper level than a literature review alone would have allowed. The second questionnaire, administered in fall 2013, was used to compare students' pre- and postinstruction understandings about key ideas, concepts, and principles introduced in the course. A qualitative analysis of responses revealed several shifts, three of which pertained directly to engineers'/scientists' relationship with "the public": (a) At the beginning, students associated engineering/science ethics with abstract rules. At the end, their understanding revealed a shift to how engineers/scientists operate in real-world contexts and, more specifically, to their relationship with the diverse publics affected by their work. (b) At the beginning, students characterized "the public" as different and separate from engineers/scientists (e.g., general population, "herds of sheep," organizations/companies). At the end, numerous students described it in relation to engineers/scientists, focusing on the power differential between the two (i.e., the public being affected by engineers/scientists but having limited control over their work). (c) At the beginning, students tended to view engineers'/scientists' interactions with the public as risky because they felt that individuals who lack proper training can misunderstand or misinterpret technical information. At the end, students added to these risks that the information communicated by engineers/scientists can sometimes itself be inaccurate, incomplete, or even deceptive. Some students also asserted that engineers/scientists should not hesitate to communicate technical information to nonexperts because the public has a "right to know" and, when treated with respect, can be a "powerful ally." These responses suggest that the ethnographic component of our class helps expand how students see engineering/science ethics and inspires them to reimagine (a) who "the public" is, (b) who they, as engineers/scientists, are, (c) what the power differential between experts and nonexperts might be, and (d) how they can relate to the publics they might one day affect in collaborative and empowering, rather than paternalistic or exploitative, ways.

Additional resources:

L2L assignment: https://www.filesanywhere.com/fs/v.aspx?v=8b6f6a895c6072aca8
2012 Syllabus: https://www.filesanywhere.com/fs/v.aspx?v=8b6f6a8b595e6fab72a2
ASEE paper 2014: www.asee.org/public/conferences/32/papers/10155/view
2012 student blogs: https://blogs.lt.vt.edu/vt5804ethicsandpublic2012/
Video excerpt of student presentation: https://vimeo.com/138734465

Humanitarian Engineering, Past and Present: A Role-Playing First-Year Course

Institution: Worcester Polytechnic Institute

Faculty/contributors: Kristin Boudreau, Laura Robinson, Leslie Dodson, David DiBiasio, Curtis Abel, John Sullivan, Glenn Gaudette, John Bergendahl, Chrysanthe Demetry, Paul Kirby, Kristin Wobbe, Joseph Cullon, Nicholas Campbell, Adam Carrier

Exemplary features: Multidisciplinary faculty; use of role playing and interactive teaching, which exemplifies the strategy of using games as a pedagogical approach

Why it's exemplary: This course is exemplary because it teaches engineering content in a complex social environment where ethical questions are part of engineering practice. Whether they role-play engineers, businessmen, scientists, or laborers, students learn and practice engineering content (fluid flow, chemical precipitation, sand filtration, water analysis) while also learning to understand the different points of view that often complicate the simplest technical solution. Debating macroethical questions (should the city install a new sewage system even though the law still allows dumping of sewage into canals and rivers?), students learn to address complex social problems with creativity, cross-cultural communication skills, and an appreciation for diverse viewpoints. The interactive format encourages engagement and deep learning, while student reflections at the end of the role-play help them examine their own views and understand the views of others. Later design projects invite students to practice what they've learned in a contemporary context.

Program description: This role-playing activity takes place over 7 weeks in the first half of a first-year general education class. Enrollment is 30–60 first-year students, predominantly but not exclusively engineering majors. The course is cotaught by faculty in humanities, social sciences, and engineering disciplines. Goals of the course are to introduce students to the multiple perspectives, disciplines, and abilities needed to solve complex, open-ended problems: the ability to identify answerable questions and to select and evaluate suitable solutions through the application of more than one discipline; to work effectively as collaborators on a team; to effectively research and use sources; to communicate clearly, effectively, and with appropriate evidence; to understand and articulate the differences in experiences of a complex problem; and to understand one's own and other people's values when they concern complex engineering problems.

Students work collaboratively in the second half of the term to propose an engineering solution to a problem of sanitation or water access; in the first half of the term, they role-play an actual 19th-century urban sanitation project. Set in Worcester, MA, in the 1890s, this game provides a complicated context of labor dissatisfaction, social inequality, rapid urbanization, and cutting-edge engineering resources and practices, simulating for students a complex engineering problem. The question of how best to mitigate the extreme pollution in the Blackstone River and Canal becomes a launching point for students to discover some of the social, environmental, and economic difficulties that complicate professional engineering practices. As they inhabit roles, conduct research using primary sources, and discuss their perspectives with other players, students learn to understand the technical and nontechnical issues deeply from a particular perspective. As they come to terms with the ethical issues and begin to understand how different values are weighted in a diverse community and what tradeoffs are made, they exercise agency as game players.

This active, immersive, interactive role-playing game includes introductory engineering and science content (e.g., as students learn about fluid flow or structure in preparation for designing a prototype sewage system or when they conduct water quality tests for chemical pollutants and bacteria). At the same time, and based on the science of both quantitative and qualitative data, students must decide where they will stand on an issue and try to persuade others. If the law doesn't require improvements to the polluted river, for instance, should a politician or city engineer recommend improvements? What about when sewage overflows into working-class neighborhoods threaten the health of a population? Does it matter that they don't speak English and cannot vote? Or that other engineering projects are being urged by industrialists? What if the city engineer has the chance to help develop a state-of-the-art sewer design? Decisions in this game have consequences, and the dilemmas are morally ambiguous rather than clear-cut and didactic. At the end of the role-play, students reflect on their decisions and discuss them with the class. Because the ethical lessons go well beyond the prescriptions that tend to dominate ethical education when considering codes of conduct and standards of ethical practice, these activities and especially the indefinite nature of many of the decisions help students develop into better, wiser engineers who are able to make their own, autonomous decisions.

Course activities include community dialogues that entail (1) scoping an engineering problem (by ascertaining what the overflow really means, whether it is a threat, and how great a threat it is); (2) deciding whether to take action (when law and some moral codes are not entirely in sync); and (3) deciding which action to take (as students present different sewage engineering designs). These goals include many of ABET's student outcomes: oral, written, and visual communication; ability to function on a multidisciplinary team; ability to analyze and interpret data; understanding of professional and ethical responsibility; and the broad education necessary to understand the impact of engineering solutions in a global, economic, environmental and societal context. This education includes the ability to conduct research: to find and evaluate a variety of sources in order to help make important decisions. The method of the course is to assign roles based on real 19th-century people and activities and objectives for each role. Students must conduct research, find potential allies, identify foils, and negotiate tradeoffs in order to realize their goals and objectives. In many cases they have a specific objective but their characters are indeterminate on some ethical issue, so they must think through their positions and later reflect on their ethical choices. Activities include combinations of research, hands-on experiments, and scientific and moral arguments using both quantitative and qualitative data. For instance, after examining 19th-century water quality sets and mapping them against city maps that reveal economic status, players deliver arguments (informed by data as well as economic, environmental, and moral values) about policy.

In the last half of the semester students work in teams to identify and design engineering solutions to water and sanitation problems in the developing world. Here they must once again consider the tradeoffs between economic strength, social well-being, and environmental sustainability. A team of faculty advisors including engineers and nonengineers works with these students as they learn to define a problem and propose a solution. For instance, a team of students might design a sanitation station (laundry, toilets) for a Namibian village that combines engineering technology for sanitary conditions with sensitivity to cultural practices, so that the women of the village are most likely to value and use the station.

Assessment information: To understand their roles and meet their objectives, students playing the game identify primary and secondary sources that were available in the 1890s and provide a list of the sources consulted. To assess their information literacy, we review these sources and analyze them for quality and variety based on assessment methods reported in a number of studies completed in recent years focusing on ways to assess information literacy skills outcomes of undergraduates (see Boudreau and Hanlan 2014). Interdisciplinary teams of faculty review student posters for engineering content. Here, students present their final projects and are evaluated according to the extent to which they have identified an answerable question and evaluated and selected a suitable solution through the application of multiple perspectives and disciplines. Communication is evaluated by reviewing essays, visual displays (posters, slideshows, graphics), and presentations for clarity, effectiveness, and sound use of evidence to support conclusions. Qualitative assessment of information literacy, ethical reasoning, and creative thinking is done by assigning and reviewing reflective essays, which provide insights into how students use and interpret these sources. This assessment was inspired by the work of researchers investigating professional competencies in engineering education through assessment of student portfolios and reflective writing (see Boudreau and Hanlan 2014). At the end of the role-playing activity, students submit an essay describing the interplay between technical and nontechnical concerns and reflecting on their ethical choices. At the end of the second term, after completing the team project, students again reflect on how their awareness of different experiences of the problem guided their problem-solving process. While we have not yet assigned these reflective essays to game players, we did assign them to the game developers (also students) who were responsible for creating authentic historical roles with complex ethical content. Here are the words of one of these students: "I needed to know more than the basic facts of my characters' identities. I was looking to learn about their ethics, their morals, and their intrinsic motivation, especially in relation to the pollution caused by the sewage in the Blackstone River. I found myself asking questions about my characters that were sometimes difficult to answer. Who were they mentally and emotionally? What was their attitude about their community? What did they want to see happen in order to fix the sewage problem? What didn't they want to see?" In a pilot study of this role-playing game, students were asked two additional questions as part of their end-of-term course evaluations: How much have you learned about ethics in this class? and How much has the work you've done in this seminar kindled your interest in thinking about the ethical dimensions of science, engineering, or business? The average score on a Likert scale of 1 (not very much) to 5 (very much) was 4 for the first question and 4.6 for the second.

Additional resources:

Bordoloi LM, Winebrake JJ. 2015. Bringing the liberal arts to engineering education. *Chronicle of Higher Education*, April 27. Available at http://chronicle.com/article/Bringing-the-Liberal-Arts-to/229671/

Boudreau K. 2015. To see the world anew: Learning engineering through a humanistic lens. *Engineering Studies* 7(2-3):206–208. Available at www.tandfonline.com/doi/full/10.1080/19378629.2015.1062506

Hanlan LR, Boudreau K. 2014. A game-based approach to information literacy and engineering in context. IEEE Frontiers in Education Conference, October 22–25, Madrid. Available at http://digitalcommons.wpi.edu/gordonlibrary-pubs/4/

The University of Virginia SEAS Senior Thesis: A Culminating Activity

Institution: University of Virginia

Faculty/contributors: All faculty in the UVa Science, Technology, and Society (STS) Program (contact: Deborah Johnson)

Exemplary features: Multiyear program interweaves technical education with ethics and STS education

Why it's exemplary: The University of Virginia School of Engineering and Applied Science (SEAS) senior thesis activity is exemplary because it challenges students to integrate social and ethical analysis with engineering by building on an understanding of the relationship between engineering, technology, and society. It is also exemplary because it provides this ethics education to every engineering student at a large research institution.

Program description: The UVa School of Engineering and Applied Science (SEAS) requires that all undergraduate students— approximately 650 students annually—complete a senior thesis in their final year. For this culminating activity students integrate their work on a technical project with research on an ethical, social, or policy issue related to the technical project. They develop and demonstrate their capacity for social analysis, ethical reasoning, and written communication, abilities cultivated during a four-year experience that includes four courses offered by the Science, Technology, and Society (STS) Program. The thesis requirement is managed by the STS Program, with much of the work done as part of two fourth-year courses. The senior thesis consists of a portfolio of documents:

- A prospectus: This document, written in STS 4500: STS and Engineering Practice, is a plan and justification for undertaking a technical research or design project and writing an STS research paper.
- A technical report: This document is written under the supervision of an engineering faculty member as a report on a research or design experience done either as part of a capstone or design course or an independent study. It is written in the language and style of the specific engineering discipline.
- An STS research paper: This document, written as part of STS 4600: The Engineer, Ethics, and Professional Responsibility under the supervision of an STS faculty member, is focused on an ethical, social, or policy issue related to the technical project.
- A sociotechnical synthesis: This document briefly describes the technical report and STS research paper and articulates the synthetic connection between the two.

These four documents are bound together, submitted as the senior thesis portfolio, and kept (in circulation) at the UVa Science and Engineering Library.

The SEAS senior thesis demonstrates the student's learning processes in integrating the technical and the social/ethical. Be it analysis of the social implications of a technology, grappling with a policy issue surrounding a particular technology, or a sociohistorical study of an engineering endeavor, students are challenged to develop a narrative about their technical work, the social/ethical implications of that work, and how the two are intertwined. The senior thesis is best understood in the context of the broader curriculum.

SEAS requires all engineering students to take four STS courses that aim to develop students' (1) competence in ethical awareness and analysis; (2) oral and written communication skills; and (3) understanding of the relationships among science, technology, and society and the implications of these relationships for engineering practice. These three goals are integrated in the course offerings culminating in the SEAS senior thesis. Hence, it is not possible to understand the significance of the thesis without explaining the full curriculum. The first course focuses on breadth, the second on depth in a particular subject, and the third and fourth courses on synthesis and integration. Ethics, especially professional ethics, is introduced in the first course, touched on in the second and third courses, and the major focus of the fourth course.

STS 1500: Science, Technology, and Contemporary Issues – This course is designed to introduce students to the relationship between engineering, technology, and society and to strengthen writing and speaking skills. Among many other things, the course provides students with an introduction to engineering ethics and the legal and social dimensions of engineering practice.

STS 2000/3000: Science and Technology in Social and Global Context – All engineering students take at least one 2000-/3000-level STS course in their second or third year. These courses examine specific social and/or ethical issues involving science and technology from humanities and social science perspectives. Students might, for example, explore technology in utopian thought or environmental policy or the history of technology. This provides depth in understanding the intertwining of engineering, science, technology, and society.

STS 4500: STS and Engineering Practice – Students write the prospectus as part of this course, which also has subject matter content, engaging students with the challenge of framing and solving engineering problems in a manner that requires attention to social dimensions. Students are introduced to STS theories and methods as a means to prepare them for their STS research paper.

STS 4600: The Engineer, Ethics, and Professional Responsibility – This course on ethical issues in engineering challenges students to analyze ethical issues in a systematic way. Much of the course is also devoted to completion of the STS research paper on an ethical, social, or policy aspect of the technical project.

STS 4500 and 4600 are taught in classes with approximately 30 students. We offer over 20 sections of each course.

The four-course curriculum is based on the premise that to effectively teach engineering ethics, students must be exposed to the subject matter more than once in their curriculum, and the more often, the better. Hence, the first course introduces ethics and the second courses address ethics, values, and decision making in a more specific area of specialization (e.g., information technology or nanotechnology). The two-semester senior-level courses allow students to do research on a topic of their choice as long as it addresses an ethical, social, or policy issue related to a technical project they are working on with an advisor in their major. A second premise of the UVa senior thesis is that it recognizes that ethical issues in engineering practice do not arise abstractly or theoretically or in a vacuum. They are embedded in social contexts and, to come to grips with these issues and figure out whether and how to take action, engineers need concepts and language with which to analyze social context. The field of STS provides concepts and methods that help to do just this. Students use STS to think about the social and technical together. This culminates in writing the STS research paper.

The four-course and senior thesis requirements fulfill at least three of ABET's student outcomes criteria: an understanding of professional and ethical responsibility, an ability to communicate effectively, and the broad education necessary to understand the impact of engineering solutions in a global and societal context. The courses also ensure that students produce a portfolio of work that can be and is used in the ABET assessment and evaluation process.

Assessment information: We know we have achieved our goals when we see our students writing and speaking competently and confidently about the social and ethical implications of technology and engineering, using language and concepts from the social sciences and humanities in relation to technologies, engineering challenges, and engineering endeavors. We use a variety of assessment resources. Because the senior thesis experience is managed through the two fourth-year STS courses, student evaluations of these courses provide short-term, immediate feedback. For longer-term feedback, we pay attention to surveys of alumni undertaken by SEAS, for example, about the influence of curriculum experiences on current work life. We rely most heavily on an examination of the STS research papers. Each year after the final thesis portfolios have been submitted, faculty who have taught the senior thesis courses engage in a joint assessment activity by examining and rating a sample of the STS research papers on a specified set of criteria. The criteria used for this assessment vary somewhat each year, but they are targeted to connect to the ABET student outcomes (mentioned above). [We use materials from this assessment during ABET reviews.] This assessment process reveals to what extent students are able to do the kind of analysis we aspire for them to do. We identify what is lacking in the lower-quality papers and what is exemplified in the best papers and use this information to guide our teaching. In addition to assessment, this activity helps the faculty to develop shared standards for grading.

Another kind of feedback on how we are doing occurs when STS faculty evaluate STS research papers for presentation at an annual SEAS event recognizing excellence in undergraduate engineering. Each year SEAS holds an Undergraduate Research and Design Symposium to celebrate students who achieve excellence in their technical research, design, and STS research. This past year approximately 30 students were nominated by their STS instructors to present at the symposium, a dozen were selected, and two received special awards.

Other signs that we are achieving our goals come from the achievements of our students. For example, this past year three students were invited to present their STS research projects at the Science and Technology Global Conference hosted at the National Academy of Sciences in April 2015. The conference allowed the students to interact with graduate students and faculty who focus exclusively on the ethical, social, and policy aspects of science and technology. Faculty also work with students who want to have their STS research papers published. In summer 2015 a student published his paper in *Intersect: The Stanford Journal of Science, Technology, & Society*.

Problem-based Learning in a Professional Ethics Course for Undergraduate Engineering Students

Institution: Georgia Institute of Technology

Faculty/contributors: Robert Kirkman

Exemplary features: Incorporation of ethics into engineering design thinking and engineering projects

Why it's exemplary: I am in the midst of a long-term project in design research, developing and refining an innovative approach to teaching stand-alone, semester-length courses in practical ethics for students in engineering degree programs at Georgia Tech. The design centers on problem-based learning (PBL): students work together in groups, with guidance from the instructor, to develop, analyze, and respond to complex, open-ended problem situations in professional practice. The goal is for students to envision a range of possible options for responding, then carefully to consider the ethical implications of each in terms of basic moral values.

Program description: The engineering ethics course fulfills the ethics requirement of many of the undergraduate engineering degree programs at Georgia Tech. Students are mainly third- and fourth-year undergraduate engineering majors. Each section of the course as I teach it is capped at 35 students. The course is offered by the School of Public Policy (SPP) under the Philosophy of Science and Technology (PHIL) designation. In the context of the SPP, instructors have considerable latitude in course design, outcomes, and requirements within broad outlines laid down in the course catalogue, although I am the one faculty member involved in this particular design project and the design is my own. I generally offer a section of engineering ethics every term with the goal of helping students to develop cognitive skills associated with moral imagination.

The learning outcomes of the course all involve developing the capacity to notice, respond to, and think about ethical values in particular, concrete, messy problem situations. A messy problem is one in which there may not be just one correct option or even just one way of understanding the problem. By the end of the term students should demonstrate improved abilities in the following areas:

Contextual Awareness: choose an appropriate scale for framing a problem situation and its implications; identify plausible opportunities for and constraints on choice and action in the situation; and connect opportunities and constraints to wider systems and institutions on which they are conditioned.

Critical Consideration: identify concrete instances of basic ethical values that are (a) in play in and (b) implicated in particular options for action in a problem situation, including values that tell for and against each option.

Theoretical Understanding: organize and connect concrete instances of basic values by appropriate use of theoretical frameworks; use appropriate terminology for each theoretical framework; draw appropriate connections among concepts within theoretical frameworks; and manage the connections among concepts between frameworks.

Three Auxiliary Outcomes: generate a variety of distinct, practicable options for responding to a problem situation, which includes reframing the situation (creativity); organize written work for ease of understanding, using clear and precise language that is accessible to a general audience (communication); and collaborate effectively with others (collaboration).

Following the PBL model, the course is structured as a kind of apprenticeship; students work together in groups, with guidance from the instructor, to acquire and use the cognitive tools of ethical inquiry and problem solving. The course is divided in three parts, the first part of which is an introduction to the tools themselves: students work from primary and secondary sources in ethics to develop a working understanding of ethical theory as a way of focusing attention on basic values, and they work through a series of short practice exercises to refine that understanding and develop a sense of the steps to be taken in ethical inquiry. In the current version of the course, the ethical theory on offer is Aristotle's virtue ethics, which lends itself quite well to the context of professional practice. While Aristotle was most interested in dispositions of character that contribute to general human flourishing, in terms of the function or characteristic activity of humans as such, I have students consider the function of engineers as professionals and the dispositions of character most appropriate to that role. I also provide the students with scaffolding, which is an important element of the PBL approach. Scaffolding is an artificial structure for focusing students' attention until they have enough experience to focus attention in those same ways on their own. The scaffolding for virtue ethics has students fill in columns picking out aspects of the experience of a situation and the response to it that are especially important for assessing the appropriateness of the response. In each of the remaining two parts of the course, groups work more independently to develop a novel problem situation that might confront a practicing engineer, analyze the context of the situation, develop at least three options for responding to it, and consider the implications of each of those responses. In developing a problem situation, groups essentially write a story in which the protagonist, a practicing engineer, comes face to face with an ethically fraught choice. Groups may adapt their stories from actual cases or from their own experience in co-ops, internships, and labs, or they may create works of fiction based on general understanding of the kinds of situations practicing engineers might face. In any case, stories are to be in the present tense and in the first person, and they should be open-ended and ethically complex, even messy. At the culmination of the project, each group presents the situation and responses to the class in a creative format (often a skit or short video) and facilitates a class discussion of the ethical implications. Each student then writes an "individual consideration," which begins with the problem situation from the student's working group and one of the options from the presentation. The student adds a new option of her or his own devising, then considers the implications of each option following the outlines of the provided scaffolding, but in paragraph rather than tabular form. The goal of the consideration is to indicate the basic values that are in play in each option, for good or for ill, without coming to conclusions.

Assessment information: Design research proceeds by an ongoing process of making incremental improvements of a course design, accompanied by a more formal assessment using qualitative data, quantitative data, or both. An important first step, though, is to arrive at a course design stable enough to lend itself to repeated rounds of assessment over a period of time. My project has just reached the point of having a stable course design. Significant revisions before the spring 2015 term and additional refinements just before the fall 2015 term have yielded a design that promises to function well for the foreseeable future with relatively minor modifications. This is to say that I have not yet begun the systematic collection of data on the effectiveness of the course design. I can offer some observations, though, that suggest the PBL approach in practical ethics is especially promising and that may provide some basis for the formal assessment still to come. When I started to implement PBL in practical ethics courses in fall 2012, the impact on student engagement was immediate and dramatic, especially compared to my previous lecture-and-discussion approach. Attendance improved markedly, and students were generally active participants in group work. The current design, with its high degree of student control over problem situations and presentation formats, seems even to make the course enjoyable for students. The one learning outcome that has been most elusive has been theoretical understanding, and many of the recent revisions to the course design have been aimed at bringing that more within students' reach. The current term is the first in which I have given over the first third of the course to preparatory readings and exercises, and early signs are promising that students will more quickly gain competence and confidence in using the scaffolding to identify basic values. Anecdote is problematic as evidence for the success of a design, but there have been a number of instances in each term in which I have used PBL when students have written to me or told me that being in my class has begun to change the way they perceive various situations in which they find themselves at school, at work, listening to the news, or even just spending time with friends. They tell me they have started noticing the values that are in play in such situations. They sometimes ask me questions aimed at helping them to clarify the ethical aspects of such situations One student even claimed I had "ruined" things by making it impossible to ignore values in everyday life!

Additional resources:

PHIL 3109: Engineering Ethics – Syllabus, fall 2015:
https://drive.google.com/file/d/0B1n5fQEuOtUxQldlLUEySURRd1U/view?usp=sharing
PHIL 3109: Engineering Ethics – Evaluation Rubric, fall 2015:
https://drive.google.com/file/d/0B1n5fQEuOtUxQTNPNGdNSDZIb0E/view?usp=sharing

Case Studies for Engineering Ethics Across the Product Life Cycle

Institution: Northeastern University

Faculty/contributors: Matthew Eckelman, Chris Bosso, John Basl, Jacqueline Isaacs

Exemplary features: Adaptability for use in secondary education; extensive collection of cases on the ethics of lifecycle impacts and sustainability

Why it's exemplary: Real-world engineering decision making involves multiple actors and, for each, ethical considerations may arise at multiple levels—personal, professional, societal, or global. Our program of case studies and educational materials is exemplary in its interdisciplinary foundation, created collectively by engineers, policy experts, business professionals, and ethicists to provide clear examples for rising engineers to appreciate ethical issues from multiple angles. Accompanying materials are rigorously assessed in the classroom by internal and external evaluators based on national educational goals and guidelines, with versions developed to suit a variety of instructional modes. Full cases are designed for university engineering students, while streamlined versions for secondary schools spread an awareness of lifecycle issues and environmental ethics early in formal education. Widespread dissemination using various media adds to national infrastructure for ethics education in engineering and environmental fields, with the goal of emphasizing societal ethics and indirect effects.

Program description: A central goal of engineering education is to provide students with an understanding of context for their designs and decisions. A common theme currently relates to the environment and public health, specifically what constitutes a fair distribution of emissions or impacts, who or what has value, and what exactly gets counted in an engineering analysis of benefits and costs. These questions can be quite effectively discussed in the context of lifecycle engineering, a design strategy that uses a "cradle-to-grave" approach to evaluate environmental and social impacts, incorporating material, energy, and economic flows as well as social and biological effects at different stages. While the use of lifecycle engineering and lifecycle assessment (LCA) tools is widespread, the modeling structure and interpretation of results involve ethical and value judgments that must be navigated carefully by the analyst and by the receiver of the results.

LCA is increasingly important in corporate and government decision making, yet there is a dearth of materials specifically designed to integrate ethics education into life cycle–oriented coursework. Our ethics education project centers on the integration of life cycle–oriented case studies in design, engineering, management, and public policy fields. Case studies are effective pedagogical tools, and particularly useful in enabling students to develop practical understanding of the ethical challenges they will face as practicing professionals by placing them in mock decision-making roles. We have conducted a thorough review of nearly 1,000 existing case studies from engineering, business, and public policy to determine common topics and themes that relate to product life cycles and environmental and health impacts. Our case studies cover current events and engineering design decisions that involve balancing local or direct effects with larger, indirect effects on society, including (a) mismanagement of industrial waste and ecological impacts from industrial accidents, specifically the inundation of several villages in Hungary from a large-volume spill of red mud, a byproduct of aluminum production (production stage); (b) the upstream implication of material selection for consumer electronics, specifically the tradeoffs between Au-coated antennas and GaIn liquid metal reconfigurable antennas, a new technology being piloted by handset manufacturers (design stage); (c) implementation of state-level policy around compact fluorescent bulbs, balancing state targets for energy efficiency, indirect emissions as a result of reducing electricity demand, and direct potential emissions of Hg during lamp breakage, both accidental and intentional (use and disposal stages); and (d) whether federal/state agencies could and should require labelling of nanomaterials in consumer products, drawing parallels with labelling efforts for pharmaceuticals and food (use and disposal stages).

Following typical case study methods, students are presented with an engineering or design decision that they need to make, accompanied by background material that provides technical, environmental, and policy context. An accompanying teaching note guides instructors with ideas for classroom instruction, emphasizing the ethical concepts that are relevant to the case and written with proper terminology in collaboration with the Ethics Institute at Northeastern and assessed by an external evaluator. Instructional materials and video footage presenting each case, as well as shorter versions for younger audiences, are being created and will be hosted at the Ethics Institute as an additional teaching resource. The creation of the case studies involved a multidisciplinary collaboration among faculty members as well as graduate students. Undergraduate students and high school teachers are assisting in the creation of versions appropriate for secondary schools. These cases have been designed as one-week modules to be incorporated in existing courses and ethics workshops.

The educational goals of this project are to:

(1) Create engaging, practical, and effective case study and workshop materials that examine ethical dimensions of LCA practice and communication, for use in courses in engineering, management, and social science;

(2) Evaluate the effectiveness of these materials through robust educational assessment while improving student learning; and

(3) Engage other secondary school and college/university instructors through demonstration and provision of instructional guides and resources to accompany the case study and workshop materials.

The overall purpose of the project is to enable engineering students and the general public to have an understanding and meaningful discussions of indirect impacts of their activities, and how to balance direct benefits and indirect impacts. Our life cycle–oriented, case-based approach to engineering ethics education will fill gaps in case study resources by addressing fundamental ethical principles and macro-ethical issues on sustainability topics, developing novel, robustly assessed educational materials where few currently exist.

Assessment information: Our case studies and workshops are being piloted in engineering, business, and public policy classrooms. We have also been working with the Center for Advancing Teaching and Learning through Research at Northeastern and our external assessment advisor, Dr. Michael Loui, to develop assessment instruments and evaluation schemes that can be used across all of the cases. We now have a scheme that covers the common ethical concepts introduced in the cases—distributive justice, weighting/balancing risks, moral status, the precautionary principle, responsibility to report, and exploitation. The evaluation scheme is based on the framework presented by the Ethical Reasoning Value rubric published by the Association of American Colleges and Universities and will be applied to five separate classes of students over the coming year in order to test learning outcomes. This project grew out of the team's experience with trying to fit existing engineering ethics cases into a life cycle–based framework. To provide a baseline for evaluating the new case studies, a review of learning assessments was carried out in spring 2015 for a mechanical/industrial engineering course, which currently uses a case study–based ethics module about the Bhopal chemical disaster, and retrospectively for the 150+ students who have passed through the course over the past several years. Review of assignments and responses informed the creation of case study teaching notes and the draft evaluation scheme. Continuing assessment will allow the project team to adjust the cases and teaching materials as necessary and add further instructional guidance where learning objectives are not being met.

Additional resources:

Devising State Policy on Compact Fluorescent Lamps: https://drive.google.com/open?id=0B79qckAaBoroZFBQcUNGRF84Z1E
MechE5645 syllabus: https://drive.google.com/open?id=0B79qckAaBoroa00weHQ1dUFmSUU

UnLecture on Software Engineering Ethics

Institution: University of Cincinnati

Faculty/contributors: Vignesh Subbian, Carla Purdy, Fred Beyette

Exemplary features: Connection to student co-op experience

UNIVERSITY OF Cincinnati

Why it's exemplary: Our program is exemplary because it (1) directly connects ethics learning to software engineering practice by methodically integrating students' professional experience (cooperative education, internship, or research) into their classroom learning, (2) promotes critical thinking of discipline-specific (computer and software engineering) ethical issues through active learning and reflection, and (3) allows for the development of new or improved perspectives on software engineering ethics.

Program description: The software engineering course is part of the 5-year undergraduate engineering curriculum, which includes 20 months of professional work experience in the form of a mandatory cooperative education (coop). The participants are computer engineering and computer science students in a core undergraduate software engineering course.

The UnLecture on Software Engineering Ethics consists of a participant-driven learning session and reflective writing components (before and after the session), all based on a structured inquiry rubric that consists of a set of carefully designed questions (included in the content section below) provided one week before the participatory learning session. Before the session, students review ("retrospect") their past coop/internship assignments, recollect details related to ethical issues and dilemmas, and document important/relevant points based on the questions in the rubric (presession reflection). During the UnLecture discussion session, students share their retrospective thoughts and learn from fellow students' cooperative education experiences. They also examine ethical practices that were realized in the course projects and assignments and analyze the differences and similarities between their experiences in industry and their learning experience from the course. Afterward (postsession reflection), the students gather and document ideas, examples, and perspectives from the session and reflect on how they will perform differently in their next co-op rotation or work assignment. The IEEE/ACM Software Engineering Code of Ethics is used as a central element for facilitating various components of the UnLecture.

The goals of the UnLecture are for students to (1) understand ethical issues and dilemmas related to software engineering in both industry and academia, (2) comprehend principles of the IEEE/ACM Software Engineering Code of Ethics and identify specific connections to software development process, and (3) increase awareness of concepts related software intellectual property (IP) and consequences of associated IP infringements.

Following is an example of the structural inquiry rubric provided to students. It should be noted that the UnLecture rubric is not necessarily an assessment rubric. It is rather intended to serve as a "blueprint" to define learning outcomes and guide students and instructor in executing activities involved in an UnLecture.

Part I: Software Engineering Ethics

1.1 What are your personal ethical principles related to (a) workplace? (b) software engineering? You may give specific examples.

1.2 What ethical questions have arisen in your professional experience? Explain how you (or the person involved) resolved the dilemma. Relate each experience to a clause in the IEEE/ACM Software Engineering Code of Ethics (include the clause #).

1.3 Pick a specific clause from one of the 8 principles in the IEEE/ACM Software Engineering Code of Ethics (include the clause #). Critique the selected clause qualitatively. Include examples, as needed. Note: Avoid using the same clause for both (1.2) and (1.3).

1.4 Were you given any kind of orientation/training, formal or informal, on ethical practices, as a part of your coop/internship? If so, please elaborate.

1.5 Explain general work/business ethics of your team/company (example: policies regarding data storage, server access, access to internet content during work, work-from-home options, etc.)

1.6 What ethical questions did you face in this course/course project and how did you resolve the dilemma?

Part II: Software Intellectual Property: Research and investigate the patent/IP war that was assigned to you and then answer the following questions:

1.7 Briefly describe the case and involved parties—e.g., who initiated the lawsuit (plaintiff) against whom (defendant), what was the plaintiff's claim?

1.8 Explain specific technical details (related to hardware, software, design, and/or name/logo) behind the claim/IP violation.

1.9 What was the outcome of the lawsuit? What is your take on the outcome (potential/favorable outcomes, if the suit is still ongoing)? Include your own perspectives.

2.0 (open-ended question) In your own opinion, what is the next "big thing" in the software industry? (i.e., which technology has the potential to revolutionize the software industry?). Identify ethical concerns related to that technology.

Given that our students have integrated cooperative education into their curriculum, UnLectures provide meaningful ways to reflect on ethical issues from both software engineering practice and classroom education.

Assessment information: The reflective writing components (pre- and postsession) are the primary tool used to assess student learning and therefore students are required to complete them individually. Based on the inquiry rubric presented above, students are given at least a week before the participatory session to complete the presession reflection; they are advised to complete the postreflection section on the day of the session, and to submit the entire report within a day after the session. This report provides qualitative evidence of each student's learning from the UnLecture, which accounted for 4% of the course grade, including attendance and participation in the learning session and the reflective writing components before and after the session. Aside from minor criticism on the amount of writing an UnLecture entails, students' reactions were highly positive and appreciative. Following are some excerpts from student feedback: (1) "The UnLectures were really fun. I really enjoyed talking about and hearing others' [ethical] perspectives from industry." (2) "This was a very informative course and I learned a lot. I could relate a lot of what I did in co-op to this course." (3) "These [questions/ethical issues discussed in UnLecture] also came up in several of my co-op interviews this year and I believe that the background information I learned in this class played a major role in some of the jobs I was offered." (4) "I thought that the UnLecture sessions were a good addition to the course." (5) "I didn't expect to write [reflective] essays for an engineering class."

Broader Applicability: The UnLecture method and rubrics can be tremendously valuable to software engineering educators, particularly at institutions that have integrated cooperative education or internships in their academic programs. With careful planning and rubric design, UnLectures on ethics can also be integrated into other electrical and computer engineering courses and courses in other engineering disciplines. Detailed methods and results have been presented and published in the Proceedings of the 2014 Annual American Society for Engineering Education (ASEE) conference.

Additional resources:

Course Schedule (please see week #12 and #13): http://secs.ceas.uc.edu/~subbiavh/EECE3093_schedule.shtml

Ethics and Engineering for Safety

Institution: Massachusetts Institute of Technology

Faculty/contributors: Nancy Leveson

Exemplary features: Connects ethics learning with engineering practice; requires consideration of difficult problems that lack clear right and wrong answers; prepares students for business-related ethics issues

Why it's exemplary: It connects students' ethics learning to engineering practice and also addresses macroethics (the broader ethical and social issues involving individual engineers and societal decisions about technology). Almost no other universities teach how to engineer for safety and integrate this education into helping students determine the responsibilities of engineers in general and themselves in particular in safeguarding human life while creating new technological artifacts. Ethics is not taught as a separate topic but as part of their responsibilities in engineering safer systems.

Program description: Over 400 graduate and undergraduate students have taken the semester-long class. I started it (25 years ago) as part of a software engineering class where students were asked to consider various ethical dilemmas that a software engineer might face and to decide how they would personally handle them. Students were first asked to create answers for themselves alone and later a class discussion was held to compare and discuss the alternatives. About 18 years ago I moved to the aerospace engineering department and the activity became a semester-long class that both considers engineering ethics related to safety and teaches how to build and operate safer systems.

The class starts with reading about risk in modern society, answering questions such as "How safe is safe enough?," and considering specific problematic ethical cases such as the Ford Pinto. Students are first given an assignment to answer questions about their own and general ethical standards and responsibility for safety in engineering, the ethics of risk-benefit analysis, and what level of risk should be "acceptable." The questions do not have a right or wrong answer but instead involve personal beliefs such as who has responsibility for safety (individual engineers? management? stockholders? government regulators?), various alternatives for controlling safety (government regulation, the legal and court systems), the incommensurability principle vs. cost-benefit analysis, what should be the role of the courts and legal system, and individual responsibility. Then the students discuss their answers in small groups (the class has gotten too large to have full class discussions) and report to the entire class on their discussions. Sometimes I organize a class debate with different people arguing the various sides of an issue. Students also read about the consequences of failures of engineering responsibility in loss of life, including a paper I wrote 30 years ago on the Therac 25 accidents, which has been reprinted in over 20 engineering ethics books, used in engineering ethics education, and even translated into Braille and sound recordings for the blind. The rest of the class is spent learning safety engineering and applying it to accident investigation and accident prevention. Students have a semester project on a real system (last semester it included power grids, automobile autonomy, the Iceland Blood Bank, air transportation systems, drones, manufacturing robots, and medical devices). In the projects, the students apply both engineering and ethical principles to the design, oversight (regulation, for example), and operation of the system. I try to take examples from the newspaper throughout the semester (unfortunately, it is not hard to find them) and we discuss them and also have occasional guest lecturers who happen to be in town. The course was originally a graduate class, but in the last two years I have taught an undergraduate version.

Assessment information: Assessment is done through written assignments, class discussions, tests, and the semester-long project. Class evaluations by the students are always quite high. Although the class satisfies no requirements and is a pure elective, it grows each year and this year had over 50 students. The students come from every department of the School of Engineering and students outside my department find out about it mostly through word of mouth.

Ethics as Philosophical History for Engineers

Institution: California Polytechnic State University

Faculty/contributors: Daniel Biezad

Exemplary features: Unique topical focus on historical and mathematics- and physics-based dilemmas that tie back to modern day ethical challenges in math, physics, and engineering; micro-insertion technique

CAL POLY
SAN LUIS OBISPO

Why it's exemplary: The requirement to provide ethics education in the engineering curriculum is being met in the Aerospace Engineering Department at the California Polytechnic State University, San Luis Obispo, by an unconventional approach that is intended to have a lasting impact on engineering graduates throughout their working career. Instead of relying solely on exposing students to a particular code of ethics, or on primarily reviewing engineering case studies of ethical situations, a topical history of philosophy and mathematics is presented in intermittent bursts of weekly storytelling that last 5 to 10 minutes with the intent of showing the evolution of ethics from antiquity to the present day. Surveys before and after the class showed that the engineering students appreciated and benefited from the historical mathematical and philosophical focus on ethics, and that they fully appreciated the significant ethical challenges they will encounter. Comments labelled this approach as both interesting and unique.

Program description: The primary goal of this effort is to complement and enhance exposure to and information about engineering ethics with philosophical history in a way that generates lasting internalized student concern about engineering ethical behavior; that is, in a manner that facilitates the development of what traditionally has been called a conscience, an inner feeling or voice viewed as acting as a guide for the rightness or wrongness of one's decisions and actions. This is done by focusing on philosophical and mathematical topics familiar to the student and relating them to the evolution of our shared morality. The topics must have two primary characteristics. First, their history must expose a positive and interesting relationship between a particular philosophy in a given era and the accompanying development of mathematics; for example, the relationship between the philosophy of Pythagoras and rational numbers in ancient Greek culture. Second, a chosen topic must be a link in the historical evolution of the ethical code that became widely accepted in Western culture after the Enlightenment; for example, the evolving concept of the number zero or of mathematical limits in parallel with the evolving primacy of scientific reasoning.

True stories and interesting cultural situations are used to highlight how prevailing norms of morality have evolved episodically in Western culture. The stories include the origins of cultural moral codes in the Axial Age; how Greek culture changed them; how they evolved into the ethics of the Enlightenment through the mathematics and philosophies of Galileo, Newton, Leibniz, and Spinoza; and finally, how today they precariously stand as ethical standards based on reason alone, presenting a serious challenge when viewed through the work of Immanuel Kant and John Locke. The intent is to illustrate a few historical highlights with which students can immediately identify; to show how difficult it has often been in the past to maintain ethical integrity; and to emphasize the serious ethical challenges that will confront students in the uncertain future. The weekly presentations cover the following ten topics and morals: (1) The irrational rationalism of Pythagoras (570–495 BCE) emphasizing the moral code of geometric harmony and proportion; (2) Zeno's Paradox and the Negation of Zero (450 BCE) with the moral code of absolute logic and perfection; (3) The female philosopher Hypatia (400 BCE) with the moral code of rationalism and pre-Enlightenment astronomy; (4) Ethics in the Dark Ages (1050–1100 CE) with the morality of religious dogma and certainty; (5) Famous women Isabella, Joan of Arc, and Catherine of Sienna (1200–1300 CE) with the morality of religious dogma and revelation; (6) The end of Byzantium and crisis in the West (1400–1600 CE) with the morality of religious dogma and Machiavellianism; (7) Cartesian mathematics and philosophy (1600–1650 CE) with "I think, therefore, I am" reasoning; (8) The calculus (Newton and Leibniz) and philosophers Spinoza, Locke, and 19th century Germans (1650–1850 CE) with Enlightenment morality under uncertainty and limits; (9) Ethical dilemmas and examples (1850–1950 CE) with the morality of secular humanism and relativist ethics; and (10) Ethics in crisis: high school shootings (current day) with morality tainted by nihilism and alienation.

The ethical challenge concludes with case studies and a general awareness of a crisis in ethics that can be stated as the following question: How can we foster and provide meaning and purpose for all individuals, no matter their talent, motivation, or status, given an increasingly materialist worldview and the individual's shrinking importance within it; that is, given an ethical worldview based on reason alone? This is the modern ethical conundrum, the moral challenge that confronts the current and probably the next generation. If the supremacy of reason—both in science and in the conduct of human affairs—is a necessary condition for a moral and ethical society in the modern world, it remains an insufficient one. In addition to a code of ethics that puts this necessary condition into words, a healthy democratic republic must also allow and promote a diverse array of belief producing individuals who are motivated to strive for excellence in all areas of life; who reject absolutisms or fanatic ideologies that lead to violence; who both accept and heed their profession's code of ethics in principle and in practice; and, just as importantly, who accept some degree of uncertainty as a fact of life and as a reality of their faith. This internal acceptance of ethical standards implants within oneself what has traditionally been called the conscience, the essence of personal integrity.

There remains the danger of overconfidence. Specifically, if one cannot allow a level of ignorance to exist in one's own views, however slight, and thereby accept the uncertainty advocated by Richard Feynman as a precondition for progress, one may abandon the difficulty of striving for virtue and either seek what is most appealing materially or succumb to what is most powerful ideologically. The caution urged by G.K. Chesterton during any search for an ethics based on reason alone should be taken to heart: "Wherever the people do not believe in something beyond the world, they will worship the world. But, above all, they will worship the strongest thing in the world."

Assessment information: Assessment of the team project is done by the clients and by the professor. We also collect individual essays in the first week about ethical issues in software design and compare them with essays done in the last week.

Additional resources:

Ethics as Philosophical History for Engineers (paper accepted at ASEE conference in Seattle, June 10, 2015): https://peer.asee.org/ethics-education-as-philosophical-history-for-engineers

Engineering a Catastrophe: Ethics for First-Year STEM

Institution: Lafayette College/Rutgers University

Faculty/contributors: Tobias Rossmann[1]

Exemplary features: Use of historical cases paired with contemporary issues/topics to examine ethics from the perspective of multiple stakeholders

LAFAYETTE COLLEGE

Why it's exemplary: Engineering a Catastrophe: Ethics for First-Year STEM is an exemplary cornerstone ethical experience because of its ability to engage students in balanced ethical and technical discussions in a diverse environment using risk benefit analysis and ethical audits to address both macro- and microethical issues. Current engineering achievements and disasters are considered in light of past failures, allowing students to both explore historical ethical decisions and see these issues echoed in current engineering challenges. Engineers benefit from the ability to take the view of a nonengineer, develop empathy, and think divergently, facilitated by the ethical discussion in environments where other majors (both STEM and non-STEM) are engaged. This exposure to ethical constructs and problem solving for first-year engineers is critical to supporting future modules of engineering ethics in later major courses that build on this solid foundation to provide vertically integrated learning.

Program description: Engineering a Catastrophe is offered as part of the one-credit Byrne Freshman Seminar program at Rutgers University. The seminar is intended to provide a broad introduction to ethics through discussions and writing assignments focused on case studies of engineering catastrophes, meeting once a week for 90 minutes, and to encourage students in college-level critical thinking skills. The main goal of the seminar is to engage first-year STEM students to discuss ethics from an engineering perspective, give them tools beyond their intuition, and assist them in their transition to college-level academic work.

Students are introduced to a risk assessment–based approach to ethical decision making. This approach incorporates basic questions of risk-benefit analysis with information on the decision makers, constraints and context, and implementation of the system. This simplified framework allows students to more easily explore complex catastrophes from multiple points of view and to draw parallels with current technological issues, with these skills significantly improving over the course of the semester. The course is described broadly to attract engineering, STEM, and nontechnical majors. In fact, the title of the course, "Engineering a Catastrophe," explicitly does not mention ethics in order to appeal to the widest audience. The course typically enrolls 20 students, 50% of which are engineers, 25% other STEM, and 25% non-STEM.

This seminar is designed to explore both the engineering and cultural implications of recent and historical disasters with examples taken from natural (e.g., levee failures, earthquake damage), engineering (nuclear power generation, aerospace), and conflict (terrorism) tragedies. Students are guided to learn and discuss which factors led to these cataclysmic events and how engineering development, public policy, and society have responded. To focus on the relevance of the course to future events, readings and discussions center on how advances in engineering both solve current problems and cause new issues and unforeseen complications.

The educational goals of the course are to understand (a) the factors that lead to an engineering catastrophe (human, economic, social, safety, environmental); (b) ethics and ethical behavior in engineering practice; and (c) how decisions throughout the engineering design and implementation processes affect the failure modes of a system. Students consider current engineering achievements in light of historical failures. A case study is used to direct the ethical discussions. However, instead of focusing on individual catastrophes, discussion topics attempt to weave several events together to create a coherent story about a single issue. For example, a typical discussion of human factors and how safety is managed in large organizations centers on how initially harmless technical or managerial decisions can grow and propagate throughout a project, eventually leading to failure. This typical topic for ethical analysis is usually framed around a single event (like the faulty oxygen cylinder on Apollo 13); but the approach in this seminar frames the topic around a single issue (the transport of pure oxygen) with a multiplicity of historical and modern examples. Short histories are given of relevant historical space and aviation events involving oxygen transport followed by a discussion of the transportation of lithium batteries. Parallels are drawn between the historical oxygen-related tragedies and current issues associated with aviation battery systems and battery transport. The description largely focuses on why these types of similar events keep occurring throughout history even though the engineering community is aware of the attendant problems.

Before the open discussion, two writing prompts are given for each topic for the students to consider individually and then in small groups. Writing prompts typically focus the students on the both the societal implications of catastrophes [A-type questions] and the personal ethical

[1] Engineering a Catastrophe: Ethics for First Year STEM was originally taught at Rutgers University. The same strategies have been employed in courses at Lafayette College.

issues [B-type questions] that a practicing engineer might face. For examples: [A] When US companies work in a global marketplace, whose laws prevail? Who takes responsibility? [B] How can safety/ethics be communicated across cultural and socioeconomic divides? [A] How can ethical and safety standards keep up with a rapidly advancing scientific forefront? [B] How do engineers best approach the unknown unknowns of new technologies used in consumer products? The success of such discussions and directed writings require the students to have reasonably well developed ethical analysis skills.

First-year students experience difficulty in objectively assessing the events leading up to these incidents with their hindsight and knowledge of the consequences. Therefore, a framework using a risk-benefit analysis (with which the students are somewhat familiar) and an ethical audit are used to give the students some constraints with which to approach their exploration. Students are instructed to evaluate hazards both in and out of the technical realm. Discussion of uncertainty in engineering design and operation is balanced with estimation of nonroutine operation, historical failures, managerial complications, and consequence potential. Hazards are then folded into a risk profile with sufficient resolution for the students to capture the most important and provocative hazards. Special detail is given to the quantification of personal/public risk and risk perception (as often the mere hint of a catastrophic risk in an engineering project can seal its fate). Finally, the original risk-benefit analyses of each catastrophe are outlined such that the students can appreciate that well-developed foresight in a large, complex system is very difficult to achieve. With additional evaluative tools students discover a greater ability to personally relate to complex ethical decisions inherent in the more complicated case studies. They find comfort in defending their risk profiles and analyses rather than relying on and upholding their own personal opinions. Using these tools, their discussions and papers present a more nuanced and enlightened approach to the discussion of the acceptability of risk. With this better understanding of risk, students have a larger appreciation for the difficulties of the ethical decision making process.

Assessment information: Assessment of this course is done through student surveys (using a typical Likert scale) and by evaluating student work from the earlier and latter parts of the semester. Students report high levels of satisfaction with the class discussions (4.6/5), their ability to consider multiple sides of an issue (4.7), and their engagement (4.7). Because introduction of first-year students to college-level academics is also an important goal, survey questions are targeted toward the students' engagement with the discussion process and their level of comfort with the new intellectual material. Students report that the course inspired them to think in new ways (4.6) and to ask questions and express ideas (4.3), engaged them with new ideas (4.8), and was a positive learning experience (4.6). Assessment of written student work is performed using a rubric that evaluated their early in-class writing assignments and their final risk assessment papers. The seminar is a one-unit course, so the number of out-of-class writing assignments is kept to a minimum. The initial writing assignment is geared toward a risk assessment analysis of cheating on exams at the college level. A short lecture in the introductory class introduces the students to the tenets of risk assessment. Students are tasked with explaining the ethical concerns by viewing the risks and benefits from many perspectives (their current standing, their future, parents, professors, school administrators, future employers, and alumni). Their papers are evaluated on the depth of their exploration of the ethics of professionalism and their ability to identify motivations of each of the stakeholders. On average, students' early understanding of ethical concepts was scored at 2.1 out of 5 and their ability to apply risk assessment tools to ethical problems was scored at 1.6 out of 5. The final risk assessment paper is a detailed examination of a catastrophe that was related to one examined during the seminar but not specifically discussed. Example subjects of final student papers are typhoons in the Philippines, postearthquake structural failures in China and Haiti, vaccinations and the swine flu pandemic, and drone aircraft. Students are asked to analyze these (potential) catastrophes in light of the historical case studies presented in class, applying the risk assessment tools developed during the seminar. Final papers are judged using the same rubric as the initial writing assignment. On average students' understanding of ethical concepts more than doubled to 4.3 out of 5 as did their ability to apply risk assessment tools to ethical problems (4.1 out of 5).

Additional resources:

Ethics for First-Year STEM: A Risk Assessment–based Approach: www.asee.org/public/conferences/56/papers/11730/view

Phenomenological Approach to Engineering Ethics Pedagogy

Institution: Michigan Technological University

Faculty/contributors: Valorie Troesch

Exemplary features: Interaction with practicing engineers on ethics issues

Why it's exemplary: My program addresses two core problems in engineering ethics pedagogy: 21st century technologies raise daunting ethical questions that require strong engagement with ethics by engineers, yet engineering students don't care much about studying ethics. I developed a phenomenology-informed approach to ethics pedagogy in which students undertake research that investigates the question, What is it to be an ethical engineer? The coursework is interactive and emphasizes ethics in real-world, lived, everyday engineering practice. Students investigate their roles as engineering citizens from macro- and microethics perspectives and develop an affective engagement with study of ethical engineering practice. In other words, they begin to care about ethics and this helps maximize their learning. Students demonstrate not only significantly improved ethical reasoning and decision-making skills but a deeper reflective understanding (versus rote knowledge) of their professional and ethical responsibilities. This approach is transferable to graduate students and is scalable and replicable.

Program description: I have learned that undergraduate engineering students who are nearing graduation are unprepared for and fearful of facing the myriad ethical challenges present in 21st century engineering practice. There is a critical gap between what students need and what we offer. While we educators are concerned with imparting ethical knowledge—codes, ethical theories, decision-making models applied to case studies—our students are concerned with understanding how they are going to fit into the world of engineering as ethically competent professionals when they make the leap from undergraduate student to practicing engineer. We must fill this gap if we expect our students to graduate with an understanding of their professional and ethical responsibilities. Based on my classroom work I've found that a phenomenological approach to engineering ethics education—where students are given the opportunity to investigate, encounter, consider, interpret, and understand the real, lived experience of what it is to be an ethical engineer—can help fill this gap.

Phenomenology is the study of human meaning from the standpoint of experience. It discloses the essences of human experiences to yield a better understanding of these experiences, to capture how it is to do or experience something and what that experience means to the persons experiencing and studying it. Importantly, phenomenology is grounded in the real, lived world of everyday human experience, not in abstract theory that seeks to explain how things are or should be. Phenomenology is particularly useful to study professional experience. Sadala and Adorno (2002), who used phenomenology to help nursing students understand the world of nursing on an isolation ward, found that this method is the most effective way for students to investigate the lived professional world because they acquire "experience in a situation where they relate to an already given world, which is out there, into which they are launched and which they will have necessarily to face" (287–288). Simply put, engineering ethics will be more meaningful to students if they study it in the context of everyday engineering work.

The two principal educational goals of my class are for students to (1) recognize the values embodied in the professional code of ethics for engineers and understand how these values influence actual personal and professional ethical decision making, and (2) have an understanding of their professional and ethical responsibilities. Students achieve these goals by conducting qualitative phenomenological and interpretive research into the question "What is it to be an ethical engineer?" Phenomenology is not a standard approach used in engineering ethics education so there were no existing models to replicate for either pedagogical or assessment purposes. I had to design and test my own model for my one-credit, 3000-level, elective course, Ethics in Engineering Design.

Students undertake three core research activities: (1) They examine their own values and the values that inform professional codes and ethical theories. Though generally not made explicit, ethical engineering practice is inherently concerned with values and value judgments. Values—even for professionals in a technical practice—are fundamental, familiar, and everywhere. Students' reflection on values brings deeper awareness of what is important to them, the priorities they choose, and how they make ethical decisions. (2) They interview practicing engineers about what it is to be an ethical engineer. These interviews are the single most influential activity undertaken by the students. The impact of this one-on-one experience cannot be reproduced in a textbook. This is where students gain a truer perspective on the ethical environment and issues they will face in practice and where many of the misconceptions about ethical engineering practice are debunked. Students routinely report that this is the activity they most dreaded but ultimately the one that was the most rewarding. (3) Students read a selection of writings presenting a broad range of perspectives on what it means to be an ethical engineer. Topics addressed include technology and the ethical engineer, sustainability and ethical engineering, roles of engineers in policy development, comparative global ethical practice and identity, and alternatives to traditional professional ethics deliberation. Students must ask how each article informs them about what it is to be an ethical engineer. It is important to review these articles each year, keep them relevant, and include a variety of perspectives. Students' final research paper draws on all this work to interpretively understand and express the essences and meaning of what it is to be an ethical engineer. There are no "right" or "wrong" answers; each student's work is personal and unique. Additionally, I meet twice with each student individually to monitor his/her progress and address questions/concerns. These meetings are instrumental in generating students' affective engagement with the class.

Notably absent from this curriculum is the traditional case study ubiquitously used to teach engineering students how to apply ethics knowledge. A serious but unheeded charge against the case study is that it creates a myth of the engineer as the "individual actor who, alone, must make the ethical decision between 'personal sacrifice' or doing nothing" (Conlon and Zandvoort 2011, p. 220). My own students express this fear but report that their research interviews usually reveal the myth is unfounded and not representative of actual engineering practice. A better approach to case studies is needed, especially when engineering problems with ethical implications cannot be solved by science alone. My students consider, for example, how ethical engineers could use rhetorical deliberation to reveal otherwise unconsidered options in these cases.

On completion, my students are affectively engaged in their work and demonstrate improved ethical reasoning skills and understanding of their professional and ethical responsibilities.

Assessment information: I assessed student learning outcomes for 3 years using both quantitative and qualitative methods. Quantitatively, I used the Defining Issues Test-2 (DIT-2), a measure of ethical reasoning skills frequently used in engineering ethics education research. It is a multiple choice test with five nonengineering-specific scenarios presenting various ethical dilemmas. My students took the test in week 1 and after week 14. In 2011 mean N2 test scores increased 23.40% (from 28.59 to 35.28); in 2013 scores increased 26.62% (from 26.82 to 33.96); and in 2014 scores increased 38.38% (from 34.08 to 47.16). These scores compare to (1) those from an NSF-funded study of ethical skills of undergraduate engineering students ("SEED" study), where mean N2 scores for Michigan Tech students and those from 17 other institutions were 29.7 and 32.4 respectively, and (2) the DIT-2 national norms for college seniors in all majors and graduate students in all majors of 36.04 and 41.33 respectively. My students usually started the course with mean test scores lower than their peers, but their scores improved significantly each year to exceed those of their engineering peers and to approximate their nonengineering peers. In 2014 their post-test scores exceeded not only their engineering and nonengineering peers but also national norms for graduate students. This increase may be attributable in part to the individual meetings I added to the curriculum in 2014. These meetings promote student affective engagement, a known contributor to improved student learning outcomes. Thus, the combination of a phenomenological approach to ethics education and attention to affective engagement enables students in this one-credit course to significantly improve their ethical reasoning skills. Although the student numbers are small (20, 20, 13), the annual improvement in results is consistent.

These students are not self-selected for their commitment to ethics. Annual surveys show that nearly all take this class because they need one credit to graduate, not because of the ethics content. I used a qualitative philosophical hermeneutic approach (which looks for evidence of understanding) to assess whether my students expressed an understanding of their professional and ethical responsibility in their final essays and found that each student has a personal view of what it is to be an ethical engineer. They are more confident about facing ethical problems because they understand that experienced people are available as resources and that ethical decisions needn't be career-ending. They appreciate and understand the complex nature of ethical decision making and that it often involves tradeoffs in values, not tidy win-win solutions. They remain ambivalent about the relationship between technology and being an ethical engineer, but they do understand that engineering practice and ethical decision making occur in and are relevant to broader social contexts beyond the laboratory. These students will be less surprised by the ethical problems they encounter in practice and better prepared than most of their peers to deliberate them. These findings were reviewed and affirmed by a panel of practicing engineers.

Additional resources:

Conlon E, Zandvoort H. 2011. Broadening ethics teaching in engineering: Beyond the individualistic approach. Science and Engineering Ethics 17(2):217–232.

Sadala MLA, Adorno RF. 2002. Phenomenology as a method to investigate the experience lived: A perspective from Husserl and Merleau-Ponty's thought. Journal of Advanced Nursing 37(3):282–293.

Teaching Engineering Ethics: A Phenomenological Approach: http://ieeexplore.ieee.org/xpl/mostRecentIssue.jsp?punumber=44

A phenomenological approach to teaching engineering ethics: http://ieeexplore.ieee.org/xpl/login.jsp?tp=&arnumber=6893434&url=http%3A%2F%2Fieeexplore.ieee.org%2Fxpls%2Fabs_all.jsp%3Farnumber%3D6893434

Corporate Social Responsibility Course

Institution: Colorado School of Mines

Faculty/contributors: Jessica Smith

Exemplary features: Coverage of a novel and important topic: critical examination of corporate social responsibility and engineers' role in it

Why it's exemplary: The social and environmental dimensions of the mining and energy industries pose vexing ethical challenges that are "wicked problems," so-called because they are (1) difficult to formulate and resolve in ways that are satisfying to all stakeholders; (2) intertwined with other major problems; and (3) too crucial to be left unaddressed. The growing significance of these industries poses special challenges for engineers from a variety of disciplines seeking to work at the intersection of corporate interests, public welfare, environmental sustainability, and professional autonomy. Yet practicing engineers report that their training in these areas occurs at work, rather than in their undergraduate study. This course addresses that gap by using social science research, lectures from practicing engineers, and real-world group projects to help students understand and question the links between engineering and social responsibility, laying the groundwork to become agents of social responsibility in corporations that must deal with wicked problems.

Program description: Corporate social responsibility (CSR), as a contested and evolving field of practice, has become the dominant framework to understand and address the social and environmental impacts of many industries, from manufacturing to pharmaceuticals. The term first arose in mining, oil, and gas companies seeking to allay public outcry over human and environmental disasters, and quickly expanded into other sectors. The field of CSR is internally varied, but policies and activities under its umbrella all share an acknowledgment that corporations must address the social and environmental impacts of their activities and improve their relationships with wider publics. CSR is not a panacea for reconciling ethics with economics, nor a disingenuous attempt to cover up the continued ills of irresponsible business practice. It is an increasingly influential suite of practices, concepts, organizations, and institutional frameworks that have transformed the ways firms organize both their internal activities and their relationships with external entities such as government agencies, activist groups, and community stakeholders.

Although CSR policies and programs shape the work done by practicing engineers, very few undergraduate educational experiences help engineering students critically investigate the strengths and limitations of CSR as a tool to manage the social and environmental impacts of their work as engineers. This upper-division course prepares students to (1) understand what CSR as a field of practice means for differently positioned actors (companies, employees, communities, etc.); (2) investigate the tensions, contradictions, and synergies in how CSR, professional codes of ethics, and personal senses of responsibility promote or hinder social and environmental well-being; and (3) identify links between the "technical" work of engineering and the "social" work of community relations to understand the sociotechnical nature of CSR. Students enroll from a variety of engineering disciplines, including mechanical (40%), petroleum and mining (30%), and environmental or civil (30%).

Case studies draw from the mining and energy industries, which pioneered CSR tools in response to critics, and class activities draw out comparisons and potential applications to other industries. To contextualize the rapid ascendance of CSR, the course begins with a lecture from a guest speaker who experienced first-hand a crisis in community acceptance of industry or what the industry terms the "social license to operate." Previous speakers have included a geologist with experience in gold mining, community development, and conflict minerals in the developing world, and a lawyer working in the area of community conflict surrounding "fracking" for oil and gas in Colorado. Course readings include social science articles that identify the key elements of CSR and compare the policies, programs, and projects enacted under this banner with other frameworks to conceptualize the relationship between industry and its publics, such as state regulation, voluntary agreements and conventions (such as those promoted by the ISO and United Nations), and legal tools such as Free, Prior and Informed Consent. It also uses ethnographic research to show how communities, especially in the developing world, use different cultural models to engage with corporations, for example as benefactors with obligations to provide financial support to poor communities rather than as "partners" or "stakeholders" who help themselves through entrepreneurial activities. The articles include cutting-edge scholarly research on CSR and detailed case studies, such as the evolution of community referendums at the controversial Marlin gold mine in Guatemala or foiled attempts at community development in the gas fields of Bangladesh. Guest lectures from industry and NGO professionals with on-the-ground experience provide opportunities for students to see CSR as a dynamic and contested field of practice that is shaped by individuals such as themselves. The goal of this section is to prepare students to think critically about the strengths and limitations of CSR to address the ethical dilemmas posed by industry, given that CSR is a voluntary set of practices, guided by private interests and organizations that sometimes intersect with government mandates and professional codes of conduct.

Students then dive deep into investigating the relationship between engineering and CSR, challenging assertions that it belongs in the "social" domain and is extraneous to technical work. They explore how the rise of offshore oil production, for example, has affected corporate-community-government relations in Africa and the North Sea, or how the design of open-pit mines engenders chronic injuries among miners. CSM alums visit to share how their work is both influenced by and contributes to their companies' community relations efforts. These perspectives examine the implications of technical design and decision making for social and environmental justice, expanding engineering ethics beyond the microscale to encompass pressing macro-level concerns.

The final week invites students to consider and share how CSR lives in their own disciplines and future careers, investigating how the particular material, social, environmental, and economic elements of nonextractive industries create different sources of conflict as well as potential tools for resolution. A series of small assignments culminate in student groups producing an original, researched stakeholder engagement strategy for a real-world engineering project. Using environmental impact assessments, social science research, news articles, and other sources, students identify, prioritize, and analyze the project's stakeholders and their needs; design methods that meet global performance standards for engaging stakeholders; and identify the place of engineering solutions in larger social responsibility efforts. They then link their project with course readings to write an essay addressing the following: Was it possible to craft a stakeholder engagement plan that fully reconciled the needs and interests of the corporation and its stakeholders? Why or why not? What does your answer to those questions suggest about the strengths and limitations of CSR? What does your experience suggest about the role engineering should play in CSR or other frameworks for corporate-community engagement?

The final project challenges students to apply course concepts to novel contexts and create new knowledge about engineering and social responsibility in relation to corporate programs, professional codes of conduct, government standards, international conventions, and community organizing. The questions, activities, and discussions throughout the course provide a foundation for future engineers to navigate the ethical challenges underlining even the most vexing of wicked problems.

Assessment information: The newness of the course precludes long-term assessment, but initial student outcomes and the growing reach of the course indicate positive results in student learning and engagement. In addition to the final project, student learning is assessed on Analytic Reading Memos, which challenge students to distill, critique, and extend the main argument of a scholarly reading; oral presentations; a synthesizing midterm essay; and an in-class debate. Progress over the course as a whole is measured through pre- and postessays in which students respond to the following questions: Do corporations have responsibilities to society? Why or why not? If you think they do, what are those responsibilities? What role does engineering play in relation to fulfilling those responsibilities? Comparing the pre- and postcourse essays reveals significant expansion in what students view as the domain of CSR; increased complexity in defining and critiquing the term; and more sophisticated understanding of its relationship with engineering. For example, the majority of students initially flag only environmental performance as a contribution of engineering to CSR, leaving aside community development, but end the course identifying how even the most minute engineering decisions impact the wider well-being of communities.

Student response to the course was overwhelmingly positive; students outlined the value of the course for their engineering careers, and one said it was the "most relevant Liberal Arts and International Studies class offered at this school." Students report that they introduce the topics and debates of the course in later ones. Perhaps the strongest testament to the course is the expansion of its core topics throughout CSM. The course links the school's Humanitarian Engineering program and its Social Justice curriculum to the school's historic strengths in the extractive industries, which were previously outside the scope of the Humanitarian Engineering program. Professor Smith gives an invited lecture on CSR each semester to Nature and Human Values, a required first-year ethics and writing course, and will lecture in the senior seminars in both Mining and Petroleum Engineering.

The success of the course has resulted in the creation of an additional upper-division course that addresses social responsibility and engineering for natural resource development in indigenous communities. The course also laid the groundwork for Smith's recent $450,000 NSF grant in the Cultivating Cultures for Ethical STEM program ("The Ethics of Extraction: Integrating Corporate Social Responsibility into Engineering Education," Award 1540298), which will ethnographically investigate how engineers working in the mining, oil, and gas industries understand and practice social responsibility and what role particular undergraduate educational experienced played in preparing them (or not) to navigate the social and environmental challenges of their professional practice. It will then use these data to integrate a critical perspective on CSR into engineering as well as social science and humanities courses at Mines, Virginia Tech, and Missouri University of Science and Technology. Finally, the course inspired the vision for the ongoing planning of a new institute at CSM dedicated to socially responsible engineering, which would be the first of its kind.

Team Ethics Assignment: Based on Engineering Student Co-op Experience

Institution: University of Wisconsin–Madison

Faculty/contributors: Laura Grossenbacher

College of Engineering
UNIVERSITY OF WISCONSIN–MADISON

Exemplary features: Use of students' own experiences in co-ops and volunteering to tackle real-life problems; use of the pedagogical approach of having students learn from their peers

Why it's exemplary: Our classes enroll engineering students at the junior level for 12 ABET-accredited engineering programs. Many of these students come into the class already having had an engineering co-op experience, and we ask them to share their stories about ethical dilemmas they have faced in the workplace. We put them in small groups (of 4–5) and they come to consensus on the most compelling, most troubling, most complex ethical dilemma; they write up the case and give a presentation to the class about it, with at least three or four options for resolving it. Then they test their options using not only the NSPE Code of Ethics but also an ethical decision-making model that includes moral tests. Finally, they explain how they would communicate their solution to necessary stakeholders.

Program description: Undergraduate students and faculty participate in these presentation sessions. Our educational goals are to ensure that students are reflecting on their work experiences in thoughtful ways and to have them articulate the ethical dilemmas that can arise in workplace contexts, solve those dilemmas in a constructive environment, work through conflict within the team, and learn to moderate a discussion about ethical issues with their peers in the class. We believe this exercise can help prepare students for the ethical challenges they will face and, ultimately, improve their leadership skills.

Team Ethics Presentation Assignment: Discuss as a team the different experiences you've all had (either at work or in an academic setting) that could be considered ethical dilemmas. Choose one that the whole team feels is worth sharing with the class. Be sure you reach consensus on the case. Prepare about a 25-minute discussion that covers the details below. Be sure each member of the team has something worthwhile and significant to say to the class as part of this assignment. Prepare to ask questions and involve the audience, and expect to be asked questions.

(1) Provide some quick background on the case so we understand who the stakeholders are and what is at stake in the case. Please leave company names out and any identifying information, but give us some generic information so we understand the purpose of the company and the roles of the stakeholders there. Describe who is involved in the decision, who might be affected, and any preexisting tensions or pressures that we need to know about with these different stakeholders. Articulate *why* it is an ethical dilemma, not just a technical problem.

(2) Lay out the facts for all sides of the problem as clearly as you can. What is the nature of the dilemma? Were any elements unknown or uncertain to different stakeholders at the time, which may have had a bearing on the dilemma?

(3) Develop a list of creative options (aim for four or five), including a few that a logical person might feel tempted to choose even though they may be recognizably unethical. (Provide us with a range of potential, realistic behaviors, not just the options you have predetermined are ethical.)

(4) Analyze each option using the following questions.

- Cost-benefit test: Would the costs of choosing this option outweigh the potential benefits, either in financial terms or abstract terms?
- Utilitarian test: Does this option do the greatest good for the greatest number of people?
- Publicity test: Would I want my choice of this option published in the newspaper?
- Reversibility test: Would I still think the choice of this option is good if I or someone I loved were one of those adversely affected by it?
- Universality test: If everyone confronted with this kind of problem were to make this kind of decision, would that produce the sort of world we would all want to live in?
- Rights Ethics test: Does this option trample on anyone's rights? If so, is there anything that can be done to mitigate the incursion on individual rights?
- Consequentialist test: Would there be any potential long-term negative impacts of this option that you can foresee?
- Professional test: What might my profession's ethics committee say about this option, particularly if they were to consult the Code of Ethics in my profession?

(5) Arrive at an ethical solution using the tests above and the NSPE Code (or any code of ethics relevant to the case). Use an explicit reference to the tests above and references to any relevant part of the NSPE Code (available at the course homepage). Did your team reach consensus on the solution, or have no consensus? Why?

(6) Does your best solution require a challenging communication of some kind? Describe your next steps, and be clear about how you would communicate your resolution of the case, and to whom.

(7) Conclude your presentation by telling us what you think we can all learn from the case. What could have been done differently that might have helped people avoid the whole

dilemma? Is there any way to prevent this sort of dilemma from happening again?

Tips on the Presentation Itself: Distribute the work evenly. Break up the work reasonably so that everyone has something valuable to say, with roughly equivalent time to say it. Time your team members and keep your own team on track. Visuals should follow our guidelines for strong visuals. Powerpoint is allowed to help your team anchor the discussion; be sure we understand the background on the case, your options, your analysis, and your solution. The maximum number of slides for this presentation is seven, including the title slide. Delivery matters—we do not want you to read to us. Talk to us and use effective emphasis and a natural style—use this experience to grow more effective as a speaker in front of the class.

Q&A: Ask useful questions of your audience at the end of your presentation, to ensure that they are thinking about your solution and to get their feedback on the solution.

Assessment information: We use a rubric to ensure that students are meeting the goals described above, based on the following: (1) Clarity of the context/background information. (2) Complexity of the options/solutions or consequences of the solutions. Are assumptions carefully analyzed? Are reasonable negative ramifications anticipated? (3) Ethical reasoning skills: What systems were used to arrive at conclusions? Was the NSPE Code appropriately applied? the Ethical Decision-Making System used, with insightful use of moral tests? (4) Further development needed for any particular idea or option? (5) Teamwork: Well organized and fluid functioning as a unit? Reasonable management of ethical dissent? (6) Class discussion moderated with fairness and strong critical thinking?

Global Engineers' Education Course

Institution: Stanford University

Faculty/contributors: Bhavna Hariharan, Sheri Sheppard, Syed Shariq

Exemplary features: Use of care ethics in engineering education and teaches learning how to listen

Why it's exemplary: This course enables students to work with a community in rural India to address local sanitation and hygiene challenges. In addition to lectures, they collaborate via Skype with community members to develop solutions. The regular connection with India exposes them to the reality of ethical challenges in engineering practice. Students learn about care ethics and how to put it in practice by developing individual "care statements," which, including the community's care statements, serve as design requirements for their prototypes. A combination of experiential learning, active reflection, interdisciplinary readings, and community interaction makes students aware of the ethical implications of engineering work and of their responsibility as engineers, but instead of feeling burdened this class offers them the discourse of care as a means to navigate and practice their ethics. The course is low-budget and has a deep impact on students who have continued to engage in research for years after the class.

Program description: Stanford University's Global Engineers' Education (GEE) course provides the opportunity for students to collaborate with an underserved community globally and conceive solutions to challenges faced by the community in ways that are safe as well as mindful of and responsive to the local economic, environmental, social, political, ethical, and cultural conditions. Engineering for underserved communities has frequently imposed solutions that have proved successful in prosperous countries but that fail to have the desired impact on impoverished communities. Local conditions, both environmental and cultural, affect the solutions and their efficacy. Attempted solutions that do not incorporate local support or take into account the aspirations of the local community do not last. This has also become apparent in the growing literature on engineering for development and social justice. GEE addresses this challenge by blurring the distinction between the student engineers in their role as solution providers and the underserved community in their role as consumers of the engineering solution. The engineering students are as much consumers as the underserved community members are designers and architects of the solution and the experience of creating it together. The aim is to educate student engineers to work with rather than for the underserved communities.

GEE currently focuses on the systematic, complex, and existential problem of lack of sanitation and hygiene facilities faced by 2.6 billion people the world over. The course addresses education, safety, and dignity while enabling better hygiene and health monitoring by making the toilet a desirable, affordable, and the preferred alternative to open defecation, starting with field sites in rural India. The course is geared toward undergraduates and has attracted students from different engineering disciplines as well as nonengineering majors. The GEE curriculum fosters collaboration through three unique elements:

- Regular video calls with experts at the partner organization in India, the Environmental Sanitation Institute (ESI), allow students to engage directly with community members.
- Readings and discussions from various disciplines encourage students to consider the complexity of the problem space of sanitation and hygiene as they prototype technologies.
- The course focuses on the idea of designing with care by inviting students and community members to express their values and goals and incorporate each of these "care statements" into the final design. Students thus have the opportunity to examine daunting concerns they may have about bridging language and cultural barriers and connecting with the harsh realities that the underserved communities experience in a nonthreatening environment. The fact that the communities are real, coupled with the regular real-time connection with them, imposes an ethical responsibility on the students and provides direct experience of real work conditions.

The GEE course has two lecture hours, a weekly Skype call with field partners in India, and weekly team meetings with the instructor. It also has a strong reflection component as students maintain daily reflection journals. The curriculum is based on Dewey's philosophical understanding of learning (as a combination of active doing, undergoing, and reflection). In class students discuss literature from different disciplines such as economics, sociology, and gender studies. They share new facts they have learned as well as what they felt the authors of the papers cared about and why. Class discussions focus on discerning the methods followed, the theories that the papers built on, and most importantly how the insights from the papers reshaped or affected the problem space. Through the readings, the students are encouraged to add detail and gradually expand the complexity of the problem space. This approach exposes students to think more broadly about not only the technical aspects of their design but also the societal, environmental, ethical, and other implications. It trains them to be mindful of the different stakeholders and an appreciation for where their views may be coming from and helps them to anticipate these differences and not be surprised by them in the future.

The class follows the Stanford process closely but is differentiated by the fact that, before commencing the designing itself, the GEE team members reflect on and articulate what each of them personally cares about in the challenges faced by the underserved community. This serves as their point of view for the remainder of the design process. It becomes a method for balancing the need to provide immediate assistance with the ability to thoughtfully create breakthrough engineering solutions collaboratively with the community. The care statements are individually created as a combination of visuals and text. The process does not require building consensus or arriving at one point that the GEE team collectively cares about; rather, individual members of the ecology are responsible for ensuring that what they care about is represented in their design solution. The ecology collectively agrees to create a solution that embodies what each member cares about. This approach ensures that the community continues to stay engaged in the process. It also prevents reducing the input received from the community to mere facts and instead ensures continuity of community engagement as they continue to share what they care about and why. By sharing stories and their lived experiences they contribute to coming up with design requirements, constraints, and ideas.

The course has served as a starting point for a sustained dialogue and inquiry into how to be a good engineer and how to navigate the complex and often burdensome ethical situations that one encounters in engineering practice. The discourse of care and reflecting on care statements has proven to be an effective means for students to persevere in their reflections and develop a personal sense of ethics that is consistent with the global ethics of engineering. The course also allows students to appreciate the importance of research in improving engineering practice. Several students from the class have continued working on their inquiry, developed research projects, and coauthored papers presented at the ASEE conference.

Assessment information: To measure the effectiveness of the curriculum, a metric called Global Preparedness Efficacy (GPE) is being developed (see link below). This metric was developed in response to measuring whether the course would satisfy ABET criterion 3h, which is "the broad education necessary to understand the impact of engineering solutions in a global, economic, environmental, and societal context." The metric is built on recognizing the challenges that can prove overwhelming when working with global communities; these include the diverse cultural, social, political, economic, and linguistic contexts and accompanying ethical dilemmas. Viewed through the lens of discontinuity theory these circumstances can be disorienting and restrict students' ability to learn. By bringing a Deweyan lens we can see these moments as opportunities for learning provided there are means to restore active engagement (active doing) by the students. The measurement scheme analyzes students' reflection journals to take note of discontinuity events and examine how many resolved and unresolved discontinuity events occurred. GPE is the ratio of resolved to total discontinuity events and reflects the ability to navigate the complexity and novelty of the problem space and to create solutions to the problem at hand consistent with the global socioeconomic, political, and cultural realities. In addition to the metric, the fact that students engage with the course contents for several years after they have taken the course is a significant indicator of having achieved the goal: Students have shared anecdotes, written conference papers, added minors to their engineering degrees, and write their undergraduate thesis on subjects that they care about, articulating how their exposure to using the discourse of care to develop a personal sense of ethics has served them in navigating their undergraduate life and studies.

Additional resources:

Developing Global Preparedness Efficacy: https://circle.ubc.ca/handle/2429/53819

Terrascope

Institution: Massachusetts Institute of Technology

Faculty/contributors: Ari Epstein, David McGee, Charles Harvey

Exemplary features: Use of alumni mentors; integration with engineering projects

Why it's exemplary: Learning about ethical practices and issues is "baked into" the program as a core component of the students' work. Alumni of Terrascope thus come to see ethical practices and issues as fundamental to any problem they take on, rather than an afterthought or external requirement. Another exemplary aspect of the program is the way it empowers students to take control of their learning process, shaping goals and problems as they proceed. Finally, the program provides students with the opportunity to work on real-world, complex problems during their first year at MIT, a time when most of their other classes focus on acquiring the tools to do great things at some future time.

Program description: Terrascope is a freshman learning community at the Massachusetts Institute of Technology in which students take on complex, real-world problems in a radically student-driven, project-based, team-oriented setting. The primary participants are first-year students, but upperclassmen continue to participate as undergraduate teaching fellows and mentors. Other participants include faculty, teaching staff, librarians, and alumni mentors. The educational goals are to prepare students to take on big problems that involve ethical, political, economic, and social factors as well as scientific and technological ones; empower students to take charge of their educational experience; give students the opportunity to do important and creative work during their first year of college; enable students to understand the social, ethical, and political contexts in which their more technical work will take place; and provide students with the tools to work in diverse teams on large projects.

In the fall semester students take Solving Complex Problems, in which they are given one big problem, as a class, and told that they have a semester to solve it. A typical problem might be "Devise a plan to provide adequate fresh water to western North America for the next century." Problems always involve issues beyond science and technology and are selected such that any solution must involve multiple tradeoffs, with no "right" or "perfect" answer. They always involve environmental questions and are real-world problems that must be addressed by society. Students form teams around different components of the problem and, with facilitation by undergraduate teaching fellows and aid as required from librarians and alumni mentors, they work on a comprehensive solution. Their first deliverable is a website that describes their solution in technical detail. Their other deliverable is a public event in which they present and defend their solution before a panel of global experts.

In the spring, students may take Design for Complex Environmental Issues, in which they split into teams to do hands-on research and development on problems related to the year's topic. Often projects involve technological solutions to specific aspects of the year's problem. At the end of the semester students present their prototypes in a public "bazaar," during which they describe their work both to members of the public and to an expert panel. This class gives students the opportunity to take part immediately in implementing solutions to the problems they have studied and also to participate in a formal design and fabrication process. [Note: Some of the papers to which links have been provided below describe an older version of a spring class, in which students developed interactive museum exhibits; for internal reasons, this version is no longer taught.] Students in the spring may also take "Terrascope Radio," in which they create a radio program for the general public about the year's topic. The format and content are up to students to decide, and shows have ranged from documentaries to magazine-style programs to radio dramas. Students learn how to use this evocative medium to communicate key aspects of the year's problem to an audience without particular technical expertise. In producing the program they also develop a deeper sense of the broader aspects of the problem and its context. Every year there is also an optional field trip to a place deeply relevant to the year's topic problem; students meet people who would be affected by their proposed solution and see details and human elements that they might previously have overlooked. The trip provides deeper, contextualized learning to complement the learning done back on campus.

Assessment information: The first measure of the program's success is the quality of the students' work. Every year the expert panel is deeply impressed by the creativity and thoroughness of the students' solution. Especially telling is the question-and-answer period of the defense, usually 2 hours (following an hour-long presentation), during which panelists grill the freshmen on both general and detailed aspects of their work. Students and panelists alike are generally amazed at the depth of knowledge and sensitivity the students have acquired in just one semester. Similarly, programs produced in Terrascope Radio have been licensed and broadcast by more than a hundred stations across the country, testifying to the effectiveness with which students have learned to communicate these important issues to the public. In addition, we conduct a detailed assessment of the students' experience every year, focusing on the degree to which the program has helped them learn to work in teams on complex problems, the progress they have made in team building and project management, and the degree to which the program has deepened their appreciation of their own potential. (Anecdotal evidence agrees with students' responses: one of us teaches a project-based, team-oriented class for sophomores and has found that he needs to be sure former Terrascope students who take the class are divided equally among teams, since they are so far ahead of their peers in group work and project management.)

Perhaps most importantly, we observe the work students do in later years, after having participated in Terrascope. They tend to be campus leaders in big projects that take on difficult societal problems, eagerly seeking out challenging issues to address. Details are given in some of the papers to which we have provided links, but examples include development of earthquake-tolerant housing that can be built with local materials in mountainous regions of Pakistan; detailed analysis of the effectiveness of MIT's recycling program; and plans for ecologically sustainable temporary housing of refugees. For these students, prepared by their Terrascope experience, ethical and societal issues are at the core of their objectives and practice, motivating and shaping the work they do.

Additional resources:

Team-Oriented, Project-Based Learning as a Path to Undergraduate Research: A Case Study:
https://www.dropbox.com/s/pzbz0c7kx2n57wi/CUR-ResearchSupportive-Ch5.pdf?dl=0

Building a Freshman-Year Foundation for Sustainability Studies: Terrascope, a Case Study:
https://www.dropbox.com/s/1yvw2o3ehdz9qkh/Sust.Sci.-Epstein%2CBras%2CBowring.pdf?dl=0

Helping Engineering and Science Students Find Their Voice: Radio Production as a Way to Enhance Students' Communication Skills and Their Competence at Placing Engineering and Science in a Broader Societal Context:
https://www.dropbox.com/s/h2jtoguwjefqdjz/TerrascopeRadio-ASEE2010.pdf?dl=0

Nature and Human Values Course

Institution: Colorado School of Mines

Faculty/contributors: Sarah Jayne Hitt, Cortney Holles, Olivia Burgess, Paula Farca, Allyce Horan, Joe Horan, Alison Lacivita, Justin Latici, Ken Osgood, Rose Pass, Eric Siegel, Jessica Smith, Jim Studholme, Seth Tucker, Sandy Woodson

Exemplary features: Multidisciplinary faculty involvement; foundational required course that is evaluated over the course of the students' education; real-world ethics problems; difficult problems that lack clear right and wrong answers

Why it's exemplary: Every freshman takes Nature and Human Values (NHV), which links personal, professional, and environmental ethics to engineering, energy, and emerging technologies. Lectures and readings by diverse experts in fields ranging from anthropology and history to nanotechnology and nuclear engineering emphasize the social, cultural, political, and moral context of engineering. Research and writing assignments require students to apply ethical theories as parts of solutions to real-world engineering cases and problems. Students use strategies of negotiation and mediation to help stakeholders make decisions about the ethical use and deployment of engineering designs and technologies. We regularly assess students' baseline knowledge of ethics and engineering in context, which they apply in their senior design and other upper-division coursework. NHV is a foundational component of CSM's Ethics across the Curriculum (EAC) initiative and was integral to an NSF-funded project called "NanoSTEP: Nano-Science, Technology, Ethics, and Policy."

Program description: Nature and Human Values gives students ethical preparation for their engineering practice by highlighting ways that new technologies and engineering feats are changing people, society, and culture; exploring the evolving definitions of nature and the environment and how they impact human interactions and occupations; and emphasizing the obligation to forge ethical solutions to debates that acknowledge the values of all stakeholders. The class stresses written and oral communication as a crucial component of professional and civic dialogue, and encourages critical reading, thinking, and conversation about engineers' specific ethical obligations as professionals and their broader moral, social, and environmental responsibilities as world citizens. Participants include eight full-time faculty in the Division of Liberal Arts and International Studies, 4–5 adjunct faculty, and the director and assistant director of the division. Each year about 1,200 students take the course, most of whom are freshmen.

NHV's educational goals are that by the end of the course, students will (1) Demonstrate understanding of major ethical theories and concepts by applying them to contemporary and recent debates on technology, resource use, and environmental issues, as well as to engineering practices; (2) Critically read and analyze arguments, accurately identify the central argument in readings, and synthesize diverse points of view; (3) Construct logical, effective, well-organized arguments whose central claims are well supported and that accurately present and adequately respond to competing arguments; (4) Successfully research topics related to engineering, ethics, and the environment, make effective use of source material in a researched paper, and correctly document sources; (5) Write clear, readable, grammatical prose developed through the process of drafting and revision; and (6) Demonstrate understanding of the impact of engineering and applied science in social and environmental contexts.

The course has its own textbook, written and edited by its faculty, containing common readings and content related to engineering, ethics, and communication. Each week, all students attend a large-group lecture and also engage in 3 hours of seminar-style learning in smaller classes. They write three papers of escalating complexity throughout the semester, using skills in summary, analysis, synthesis, argumentation, and research, which culminate in the writing and presentation of a mediated solution to an unresolved debate regarding engineering and ethics. Students take a common final exam that tests their understanding of and ability to apply ethical theories in context.

Assessment information: The final paper grades and exam allow us to determine whether students (a) are able to apply ethical theories to real-life situations and (b) understand the broader social, environmental, and cultural contexts of engineering ethics. We adjust lectures and readings according to their performance on these measures. Each semester, NHV students are also more broadly assessed in six categories: Application of Ethical Concepts; Critical Thinking and Reading; Constructing an Argument; Research; Writing and Mechanics; and Engineering in Context. Students are ranked from 1 (Lacking) to 4 (Advanced) for each of these outcomes, which we use to inform our curricular development of this foundational 100-level course. The NHV assessment rubric is then compared to those for our division's 200- and 400-level courses, helping us to see the development of skills across a student's entire educational experience. Data from spring 2012 through fall 2014 show a steady increase in skill development across all NHV outcomes, with students demonstrating the highest performance in the categories of "Engineering in Context" (3.24 avg. out of 4) and "Application of Ethical Concepts" (3.06 avg. out of 4).

Additional resources:

NHV textbook: www.hmpublishing.com/featured-titles/english/nature-and-human-values.html
NanoSTEP poster presentation: https://dspace.library.colostate.edu/handle/11124/16996

Ethics Activities in the Civil Engineering Curriculum at the United States Coast Guard Academy

Institution: United States Coast Guard Academy

Faculty/contributors: Hudson Jackson, Kassim Tarhini, Corinna Fleischmann, Elizabeth Nakagawa

Exemplary features: Deeply embedded ethics education that is integrated through a multiyear program

Why it's exemplary: The Civil Engineering Program at the United States Coast Guard Academy (USCGA) fosters ethical leader development and global awareness through a breadth of required core courses in the humanities, science, engineering, mathematics, professional maritime studies, organizational behavior, management, leadership, and law. Civil Engineering faculty, guided by our ABET assessment framework, advance student development in ethics, global, and cross-cultural issues that are tied specifically to the civil engineering profession through assignments and other curricular experiences that are regularly assessed and improved. Leadership and ethical development are cornerstones of the USCGA education and the civil engineering faculty, like all faculty across campus, are charged with ensuring that upon graduation, each student has developed into a leader of character. The combination of core courses, major-specific engineering courses, and cocurricular activities provides students with opportunities to develop leadership and professional ethical conduct required for engineering practice and service as Coast Guard officers.

Program description: During their sophomore year, civil engineering students take the Leadership and Organizational Behavior (3 credits) course in which they are exposed to fundamental leadership and management concepts. Some of the concepts discussed include values and ethics, personality, self-awareness, working in teams, motivation, and setting a vision, with particular emphasis on practical leadership implications. As juniors, civil engineering students take required core courses such as Morals and Ethics (3 credits) and Criminal Justice (3 credits). As seniors, they study Maritime Law Enforcement (3 credits). The Morals and Ethics course includes two main components: (1) ethical theories, both historical and contemporary, with arguments for and against them; and (2) applied ethics, both in general and using case studies in a specific field. Throughout the semester, students examine a range of philosophical views about what makes actions right or wrong, characters good or bad, to develop their decision-making abilities, their own moral voice, and an appreciation for the place of reasoned argument in the treatment of ethical problems. Students also study and explore basic legal concepts in Criminal Justice and Maritime Law Enforcement, learning specifically about the US civilian and military criminal justice system and legal issues associated with the Coast Guard's law enforcement mission in the maritime environment. Ethical and global issues are also progressively woven into the major-specific civil engineering courses. Some examples of how professional ethics are emphasized throughout the civil engineering curriculum are highlighted below:

- Case studies, practical examples, or demonstrations are used where appropriate. For example, in Structural Analysis the instructor developed a "professional practice moment" in which students take turns presenting a current event (that includes ethical conduct) related to structural engineering during the last 5 minutes of each class.
- In the Geotechnical Engineering Design course, students engage in forensic investigation and evaluation by reviewing four case studies, one of which involves an ethical dilemma. Case studies provide opportunities for students to make the connection between theory and real-life application of engineering principles and concepts.
- In Environmental Engineering I, for a case study involving exceedance of pollutant limits, students evaluate the situation from multiple perspectives and relate the issues to the Engineers Code of Ethics. Their progress is evaluated using a rubric linked to performance indicators. Students also research and prepare presentations for the class on Superfund sites around the country to develop an understanding of various problems and remediation technologies as well as the legal, ethical, and societal issues involved in identifying and cleaning up hazardous waste sites. Coverage of professional ethics in civil engineering is provided in detail in the Civil Engineering Design course. Students in this capstone design course apply knowledge from a broad range of technical, managerial, and humanities coursework to develop solutions that consider the economic, sociopolitical, ethical, and environmental aspects of real-world problems. They produce engineering calculations, construction drawings, project schedules, cost estimates, and other necessary project-specific documents, and then communicate the results of their capstone project via a final report and presentation to their client. Major components of the course are the preparation of leadership essays as well as research and presentation of an ethical scenario from ASCE's Question of Ethics case study archive. The case studies are related to the seven canons of the ASCE Code of Ethics. Each group has 15 minutes to present its Ethics Case Study and the team facilitates a short in-class discussion. The objective is to present relevant engineering ethical situations in the classroom to stimulate discussion of the ASCE Code of Ethics and critical thinking. Students select one of the ASCE Code of Ethics canons, research, identify, and

review relevant case studies, and present their findings to the class. Following are samples of 2015 case study presentation topics:

- Ensuring the safety, health, and welfare of the public, investigated in reference to ASCE Canon 1: "Engineers shall hold paramount the safety, health, and welfare of the public and shall strive to comply with the principles of sustainable development in the performance of their professional duties."
- An engineer's misrepresentation of credentials or dishonesty, in reference to ASCE Canon 2: "Engineers shall perform services only in areas of their competence."
- Engineers who gave false geotechnical information, ASCE Canon 3: "Engineers shall issue public statements only in an objective and truthful manner."
- The proper use of professional credentials, ASCE Canon 4: "Engineers shall act in professional matters for each employer or client as faithful agents or trustees, and shall avoid conflicts of interest" and ASCE Canon 5: "Engineers shall build their professional reputation on the merit of their services and shall not compete unfairly with others."
- Fraud, ASCE Canon 6: "Engineers shall act in such a manner as to uphold and enhance the honor, integrity, and dignity of the engineering profession and shall act with zero tolerance for bribery, fraud, and corruption."
- Employer's responsibility to employees, ASCE Canon 7: "Engineers shall continue their professional development throughout their careers, and shall provide opportunities for the professional development of those engineers under their supervision."

Assessment information: USCGA has established a set of shared-learning outcomes (for all academic programs) that include leadership abilities; personal and professional qualities; the ability to acquire, integrate, and expand knowledge; effective communication; and the ability to think critically. The shared-learning outcomes are aligned with the ABET Student Outcomes, with specifically developed performance indicators related to ethics. Faculty members have created assignments and rubrics to assess student progress and improve student development in professional ethics for each performance indicator. By integrating professional ethics development and assessment in the existing civil engineering assessment model, faculty have successfully threaded this competency into the curriculum using a sustainable and effective framework. For example, the performance indicators for two ABET student outcomes, 3f and 3h, are used to assess ethics and professional issues in the civil engineering curriculum. ABET 3f, "an understanding of professional and ethical responsibility," is evaluated by the following specific performance indicators:

- 3f-1: "articulate importance of professional code of ethics"
- 3f-2: "identify ethical dilemmas and propose ethical solutions in accordance with professional code of ethics."

ABET 3h, "the broad education necessary to understand the impact of engineering solutions in global, economic, environmental, and societal contexts," is addressed with two performance indicators:

- 3h-1: "explain the economic, social, and global aspects of engineering solutions"
- 3h-2: "discuss the environmental implication of engineering solutions."

Faculty members have crafted assignments and rubrics related to these performance indicators to ensure student development in ethical and global issues relating to civil engineering. Thresholds and performance targets were established for the successful achievement of the performance indicators, with different performance targets for exams and nonexam activities (e.g., projects, homework, reports, technical paper, oral presentations). Students are considered to have demonstrated satisfactory achievement of a performance indicator if their score (grade on a particular assessment tool) meets or exceeds 70%. A course is classified as producing satisfactory student achievement on a performance indicator if it meets one or both of the following performance targets:

- Exams: At least 70% of students must exceed the performance indicator score of 70% (C grade).
- Nonexam assignments: At least 85% of students must exceed the performance indicator score of 70% (C grade).

This well-established ABET assessment system is used to evaluate student progress throughout the academic year and monitored at the end of course review, when assessment data on student performance are discussed for each course. To ensure continuous improvements, recommendations are documented for implementation during the next cycle of course offerings. Graduates of USCGA receive a degree and a commission as a Coast Guard officer: We are preparing students to provide engineering expertise while serving their mandatory 5-year commitment to the Coast Guard, and their ethics and leadership are continually service tested for a minimum of 5 years after graduation.

Additional resources:

American Society of Civil Engineers Code of Ethics: www.asce.org/code_of_ethics/

Multiyear Engineering Ethics Case Study Approach

Northeastern University

Institution: Northeastern University

Faculty/contributors: Daniel Saulnier, Bob Tillman, Tricia Lenihan

Exemplary features: Integration with co-op activities; ethics embedded in a multiyear required engineering program; use of real-world cases; strong evidence of success based on evaluation of learning

Why it's exemplary: This program is exemplary because (1) It spans multiple years, from the students' second year (before their first cooperative education work experience) to their fourth year (before their third co-op work experience). (2) It is interactive, driven almost entirely by case studies. Students wrestle with ethical concepts as if they were the engineers facing each dilemma, learning strategies to recognize and weigh competing interests, identify their own biases, and anticipate the consequences of proposed courses of action. (3) It connects to engineering practice. Lectures and discussions are led by faculty members who have many years of practicing civil engineering experience. The case study discussions during the students' fourth year draw heavily on their knowledge of actual industry practice from their co-op experiences.

Program description: Our ethics education program is required for all civil engineering undergraduate students. The department typically graduates 80 to 100 civil engineering students per year. Although student demographics change over time, this past spring semester our undergraduate population numbered 468 students, of which 34% were women and 18% international students. The two faculty who developed this activity have over 15 years of consulting engineering experience, and thus bring considerable professional and personal insights to this program. Through their professional contacts, they are able to draw on the case studies and perspectives of current practicing engineers, including many of the program's alumni and co-op employers, who understand the program and its goals. This allows these practitioner partners to shape their contributions to the program.

The goal of our ethics education program is to ensure that students develop responsible professional behavior for their engineering careers and are prepared to recognize when situations may require ethical assessment. They practice navigating the often tense and uncertain human climate surrounding ethical dilemmas and learn that, with honesty and creativity, solutions can be developed that uphold the health, safety, and welfare of the public and the environment, the honor of the profession, responsibilities to firm and client, and their own careers. One of its student learning outcomes of the Accreditation Board for Engineering and Technology (ABET) is "an understanding of professional and ethical responsibility." As an ABET-accredited program, our program strives to meet this outcome via our innovative case-based approach.

- A procession of case studies is presented, discussed, and in some cases used for written assignments. This gets the students talking about the issues and builds their capacity and confidence in identifying early-stage ethical conflicts and determining an appropriate response and future course of action.
- Cases are selected by the instructor to facilitate evaluation of different parts of the Codes of Ethics of both the American Society of Civil Engineers and the National Society of Professional Engineers. The selection of cases allows students to wrestle with sometimes conflicting requirements in the codes. Some cases have clean outcomes, others don't. The ones without definitive "answers" help students understand the often ambiguous nature of ethics dilemmas, an understanding that informs their personal ethic, analysis and prioritizing of inputs, and consideration of outcomes of alternative courses of action. The process also reinforces their engineering technical problem-solving skills and process and teaches them that creativity often leads to better outcomes, while obvious answers often have hidden drawbacks.
- Students are often frustrated by the lack of truly "right" answers like those in the majority of their other engineering classes. Over time, they come to appreciate the ambiguity of the situations studied and the importance of thoughtful, creative thinking with full consideration of outcomes.
- Co-op employment provides a deep apprenticeship experience for our students. Some will encounter ethical dilemmas on the job, while for others the case study method provides a virtual apprenticeship experience. The students wrestle with situations in the relative safety of the classroom, while interacting with their peers (who have all had different co-op experiences) and with their instructor, who has worked in the field as a civil engineer.
- The ethics concepts are revisited multiple times in two courses (in years 2 and 4 of the degree program), reinforcing the students' ethics exposure and learning. Their understanding and appreciation of the concepts mature, resulting in greater retention of the fundamentals as well as a higher level of reflection as they progress from the first to the second course.

Assessment information: The success of our program is assessed based on the students' anonymous evaluations of the junior-year course and their scores on an independent test. The teaching evaluation numerical scores are well above average, and associated comments are positive and support our assertion that the students can adapt to ambiguity, benefit from the case study framework, and form linkages between the classroom activities, co-op work experiences, and their future careers. In the past three years, the numerical score for the course evaluation question "The in-class discussions and activities helped me to learn" averaged 4.55/5.00 for the junior-year course, compared to 4.13 and 4.15 average scores for all civil engineering courses and all university courses, respectively. The junior-year course in our ethics program has a significantly higher rating for this and other metrics.

In addition to the numerical metrics, the following comments are typical of end-of-semester course evaluations:

- "The course provided us a great opportunity to have a better understanding of our careers in the future. It is more like a training class than a lecture, which is really good for engineering students."
- "The case study approach was very useful and brought to light how many different ways problems can be viewed."
- "Probably one of the most useful classes I've taken. I feel like I actually have a resource to go to and useful lessons learned that I can apply to real life."
- "The discussions were all good, thought provoking, and kept the class involved. The group activities were fun and made the classwork relevant. Important information on how to handle problems in the workplace."
- "This class was very helpful and informative in regards to how to best handle and go about dealing with future problems and ethical decisions we will encounter later in our careers. The case studies and examples were particularly helpful. Overall this was a very effective and positive class."

For the second method of evaluation, we consider the Fundamentals of Engineering (FE) examination, which is the first stage assessment for an engineer's certification as a licensed Professional Engineer. The FE exam includes several questions on the combined topic of ethics and business practices, and a separate score in this topic is provided as part of the institutional reporting. While no examination can accurately measure an engineer's capacity for ethical behavior in confronting real-world problems, historic data from this section of the FE exam provide an independent assessment of our students' aptitude for thinking through ethical dilemmas and applying rules of professional conduct. During the period from October 2005 to October 2013 (until the recent change to a computer-based examination), 400 Northeastern University civil engineering majors took the FE exam (about half of the students who graduated during this time period) and scored 1.6% higher than the national average. On the Ethics and Professional Practice section, however, our students scored 4.7% higher than the national average.

PRIME Ethics: Purdue's Reflective & Interactive Modules for Engineering Ethics

Institution: Purdue University

PURDUE UNIVERSITY

Faculty/contributors: Andrew Brightman; Jonathan Beever, University of Central Florida; Justin Hess, Andrew Iliadis, Lorraine Kisselburgh, Matthew Krane, Michael Loui, Carla Zoltowski

Exemplary features: Pedagogical design that is transferrable and reproducible; progressive learning design and approach to teaching ethical reasoning

Why it's exemplary: Our multidisciplinary team of engineering, communication, and ethics educators has developed an innovative, interactive learning system for enhancing students' ethical reasoning skills as well as their satisfaction and engagement with engineering ethics education. We first addressed the need for enhancing ethical reasoning by developing a pedagogical framework of Scaffolded, Interactive, and Reflective Analysis (SIRA) that extends beyond case-based analyses. Second, we created a coherent framework for ethical reasoning applicable to engineering by articulating a principle-based approach, Reflexive Principlism. Third, to better engage students in ethics education we developed four learning modules, each deliverable in a hybrid format for stand-alone course or embedded curricular application. Additionally, we developed an Ethics Transfer Case tool to assess students' transfer of ethical reasoning. To disseminate this work, we have published several articles based on our research findings and have begun sharing these modules and learning system with ethics educators for testing in their institutions.

Program description: For an engineer to design, practice, or lead ethically, individually or in a team, she must have competence in ethical reasoning skills, especially in light of increasingly complex social and ethical issues facing engineering. Our interdisciplinary team has developed and assessed an innovative approach and a series of interactive learning modules for enhancing the ethical reasoning skills of engineering students. We have refined and tested this pedagogical and theoretical approach to ethical decision making through multiple iterations (2012–2015) with over 60 students (senior undergraduate and graduate students, and practicing engineers) from various backgrounds. Our system of five interactive, multimedia learning modules is designed to both enhance students' satisfaction and engagement with ethics and develop effective ethical reasoning skills. The student learning objectives framing this learning system are (1) Identify and describe ethical issues in the context of historical and developing technology and engineering practice; (2) Follow a structured, interactive, iterative reasoning process to reach a supported decision in response to complex ethical deliberations; and (3) Reflect on their ethical reasoning process over multiple case studies to reevaluate the coherence between the principles, codes, and theories involved in any given case.

The educational research goals of the project center around two core questions: (1) What is the impact of this learning system on the development of students' ethical reasoning, and their satisfaction and engagement with engineering ethics education? (2) What components of this learning system contribute to change in students' ethical reasoning ability and to their satisfaction and engagement? Each of the five modules in the learning system challenges students to move through six stages of reflective analysis. Collectively, the varied cases we have developed expose students to diverse stakeholder perspectives, conflicting value claims, and contemporary ethical problems. The first module teaches the foundational Reflexive Principlism approach to ethical decision making that students apply to all subsequent cases. The next four case-based modules include a historical disaster (Kansas City Skywalk), two cases evaluating emergent medical technologies (pediatric heart valve distribution and diagnostic device development), and a novel approach to the 2010 Deepwater Horizon oil spill. Together the cases explore a range of ethical questions focusing on the specification and balancing of the principles of respect for autonomy, nonmaleficence, beneficence, and justice. The diverse range of ethical, epistemic, social, and systemic issues encountered throughout these cases has proven particularly impactful for enhancing engineering students' ethical reasoning skills.

We designed, delivered, and tested this learning system in several teaching modes, from residential courses to hybrid in-class/online to a primarily asynchronous online format. In all modes self-paced individual learning offered in a multimedia context is complemented with highly interactive small-group discussions. For the multimedia context we partnered with an innovative educational media company to build an interactive system of integrated resources embedded in actual student deliberations. Students continue deliberations in small groups of 4–5 to attempt to resolve a complex ethical issue.

The staged process of this pedagogical approach is as follows: (1) Establishing Knowledge: Exposure to the case context, scenario, and facts. (2) Perspective-Taking: Individually investigating multiple stakeholders' perspectives. (3) Compare & Contrast: Juxtaposition of student and stakeholder perspectives. (4) Inducing Conflict: Evaluation of expert (technical and ethical) opinions. (5) Justification & Decision Making: Consensus building in a small group case report determining and justifying the most ethical course of action. (6) Reflection and Reflectivity: Reflection on the balancing and application of principles. The epistemic and ethical complexity of cases increases as students work through each stage in each module. The six-stage structure is scaffolded, with higher levels of supportive materials in the earlier stages to assist students in gaining knowledge and confidence in their responses and ethical reasoning ability. The direct role of the instructor shifts from content expert to facilitating coach as the module progresses to discussion and analysis of more complex ethical issues. The final stage of meta-reflection challenges students to reflect on what they have learned and how their ethical reasoning process developed throughout the case. The final stages of the module challenge the student to higher levels of ethical reasoning consistent with those measured by validated ethical reasoning assessment instruments.

As the foundation to the SIRA pedagogical system, we formulated Reflexive Principlism, an ethical reasoning approach that is particularly applicable in engineering. It leads the decision maker to internalize a reflective and iterative process of specification, balancing, and justification of four core ethical principles—beneficence, nonmaleficence, justice, and respect for autonomy—in the context of specific case constraints, much like an engineering design process. Reflexive Principlism also addresses a pressing need in engineering ethics for a coherent ethical reasoning approach that is applicable to complex cases in an engineering context. This approach provides structure to ethical reasoning while allowing the flexibility for adaptation to varying contexts through specification and balancing of the principles. As an example, in the context of the Deepwater Horizon case study, when considering the ethicality of deeper and riskier drilling in the Gulf of Mexico, Reflexive Principlism challenges students to integrate stakeholder perspectives (e.g., of BP executives, local business owners, marine life) in their decision-making process; this adds richer specification to the principles in the case context.

Last, we developed and validated an Ethics Transfer Case tool, an innovative rubric-based assessment that evaluates students' transfer of the Reflexive Principlism approach to ethical issues beyond the course. The tool evaluates ethical reasoning along four core components of Reflexive Principlism: (1) identification and implications of the four ethical principles, (2) specification of where, when, how, by what means, and to whom the principles apply, (3) justification, or coherence between the ethical decision, the principles, and codes, and (4) reflectivity, the conscious deliberation on the process of reasoning and decision outcomes.

The PRIME Ethics learning system develops ethical reasoning skill using complex, realistic ethical cases that address both micro- and macroethical issues relevant to engineering practice and professional leadership.

Assessment information: We have continually refined and evaluated our PRIME Ethics learning system using a strategy of both quantitative and qualitative instruments to assess students' ethical reasoning skills and their satisfaction and engagement with engineering ethics education. To assess impact on students' ethical reasoning development we triangulated results among three quantitative assessment measures: (a) the well-established and regularly applied Defining Issues Test-2 (DIT2), (b) the newly developed, engineering-specific, moral development assessment tool, the Engineering Ethical Reasoning Instrument (EERI), and (c) our novel Ethics Transfer Case method. The DIT2 and EERI assessment tools measure developmental stages of ethical reasoning (based on Kohlberg's schemas) in a general and engineering context, respectively. Higher scores indicate a greater tendency toward postconventional thinking. Analyzing changes in the pre- and postcourse measures with the EERI taken by more than 60 students indicated significant increases in their ethical reasoning levels. Similar but less significantly positive changes were observed with the pre- and postmeasures with the DIT2. To provide a more granular assessment of the specific elements of ethical reasoning changes in students, we used our Ethics Transfer Case tool for the three most recent semesters. Initial evaluation of differences between pre- and postcourse scores indicated a significant increase in students' ethical reasoning using Reflexive Principlism, specifically along the components of identification, specification, and justification; however, reflectivity indicated a slight, albeit nonsignificant, increase. To assess the impact on students' satisfaction and engagement with engineering ethics education, we used a mixed methods approach with quantitative and qualitative measures: (a) a subset of items extracted from the Student Engineering Ethical Development survey to assess satisfaction and engagement; (b) a new survey instrument to assess the efficacy of our SIRA pedagogical approach; (c) a quantitative assessment of components perceived to be most effective by students along dimensions of engagement, providing new information, understanding ethics, developing critical thinking, and guiding decision making; and (d) a semistructured interview with students at the end of the course. Preliminary findings indicate that students' satisfaction with their ethics education increased in all measures after completing the learning system. Two components were repeatedly ranked most effective: (1) multimedia case videos were highly effective for engaging students and providing new information, and (2) videos of interactive student deliberations were most important to understanding ethics, developing critical thinking, and guiding decision making.

These findings provide empirical support for the efficacy of Reflexive Principlism combined with a SIRA pedagogical framework as an innovative approach to successfully engage engineering students in ethics education and enhance their ethical reasoning skills. The PRIME Ethics learning system contains highly interactive media and deliberations that encourage active engagement with learning, uses complex ethical cases that connect directly to engineering practice addressing both micro- and macroethical issues, provides an innovative theoretical approach and structure to enhance the level of ethical reasoning, and can be delivered in a hybrid online and in-class format as a stand-alone course or embedded in a curriculum.

Additional resources:

PRIME Ethics: https://engineering.purdue.edu/BME/PRIMEEthics

NanoTRA: Texas Regional Alliance to Foster Nanotechnology Environment, Health, and Safety Awareness in Tomorrow's Engineering and Technology Leaders

TEXAS STATE UNIVERSITY

The rising STAR *of Texas*

Institution: Texas State University

Faculty/contributors: Craig Hanks, Jitendra Tate, Dominick Fazarro, University of Texas at Tyler; Walt Trybula, Texas State University and the Trybula Foundation; Robert McClean, Satyajit Dutta; Fritz Alhoff, Western Michigan University

Exemplary features: Collaboration with academia and industry; multidisciplinary and multi-institutional faculty collaboration; integration of ethics content in both technical and nontechnical courses

Why it's exemplary: The extent of this effort is exemplary: We developed, implemented, and assessed two modular courses that include societal, ethical, environmental, health, and safety issues related to nanotechnology for undergraduates in engineering and engineering technology. The courses were developed in consultation with leaders from academia and industry who have expertise in mechanical and manufacturing engineering, civil engineering, electrical engineering, industrial education and technology, physics, biology, philosophy, and ethics. An important goal of the project is to recruit, engage, prepare, and encourage students from traditionally underrepresented groups to careers in science and engineering, with a focus on nanotechnology. Texas State University, a Hispanic serving institution (HSI), and the University of Texas at Tyler (UT Tyler), whose student population is 60% women, collaborated on the project. When fully deployed, we had two online courses and modules infused in 18 face-to-face courses for the first through fourth years.

Program description: A multidisciplinary team conducted the work. Members of the development and implementation team brought experience in industry, policy, and academia and expertise in mechanical and manufacturing engineering, civil engineering, electrical engineering, industrial education and technology, physics, biology, philosophy, and ethics. They are from three universities: Texas State, UT-Tyler, and Western Michigan. The implementation team includes all team members affiliated with Texas State University and UT-Tyler as well as some additional teaching faculty at Texas State. Two full-semester online courses are offered at UT-Tyler, each with the same instructor. At Texas State, project modules have been infused in 18 courses in the existing curriculum, at all levels from first to last semester, and taught by 13 faculty members.

Our central goal is to help foster ethical awareness, broadly understood to include safety, health, environmental, and social dimensions, in the next generation of engineers. We focus on emerging technologies, especially nanotechnology. Another goal is to recruit, engage, prepare, and encourage students from traditionally underrepresented groups to careers in science and engineering, with a focus on nanotechnology. A third goal is to keep students engaged: research suggests that many students—enough to stem the shortfall of US engineers—who originally intend to pursue science or engineering switch to nonscientific fields.

Our approach differs from most previous NSF-funded projects addressing social and ethical dimensions of nanotechnology in that we are seamlessly infusing modules into existing courses across the curriculum. Curricular modules are infused in nontechnical introductory courses, including a required core course in philosophy, and in technical courses from sophomore through senior level. Students engage the material in multiple contexts and through multiple methods. The central ethics modules assist students in developing moral creativity, moral judgment, and moral sensitivity. These are characteristics of morally responsible professionals and necessary for navigating situations where existing rules are silent, unclear, or require interpretation. As Mike Martin argues, moral creativity is significant in science and engineering, and developing moral creativity not only supports morally responsible work but leads to more creative and better science and technology. Moral sensitivity is a precondition of moral reasoning and action. One must recognize the possibility of moral dimensions to a situation before one can evaluate, judge, and act.

To develop moral creativity, moral judgment, and moral sensitivity and help students stay engaged with the learning process and education more generally, the project team uses multiple and varied teaching approaches. Traditional lectures are punctuated by short videos and film clips, question and answer discussion, and short individual and group assignments, with an emphasis on active learning and integrated lessons, simulations, or projects that show the relationship of concepts to the real world. One example of active learning is researching and writing case studies. As part of the mandatory philosophy course, in a special section for engineering and engineering technology students, groups of students select a topic and explore the ethical dimensions by developing a case study. Students hone scholarly research skills, including critical engagement with peer-reviewed publications, practice working in groups, create presentations, and draft articles explaining the technical and ethical dimensions of their topic to the public. The best work is submitted to Wikipedia to contribute, if approved, to public knowledge. Analysis of case studies is part of many of the modules and central to the work of the second set of modules.

The first set of eight course modules, at the freshman/sophomore level, introduces students to nanotechnology, nanomaterials and manufacturing, national security implications, and societal and ethical issues of nanotechnology. After completing this course students will be able to (a) understand the ethical and societal impact of nanotechnology, (b) understand fundamental concepts in sustainable nanotechnology, and (c) understand the nature and development of nanotechnology. The modules introduce a method for ethical reasoning that is modeled on the design process to help engineering students think about ethical problem solving as similar in structure to engineering problem solving. The second set of nine modules, an upper-level course, addresses ethical, health, and environmental risks of nanotechnology. After completing this course, students will understand (a) the health and environmental risks of nanotechnology, (b) how to work in a group and conduct systematic research to write a group-based term paper on case studies and/or research topic, and (c) approaches to assessing lifecycle risk assessment of nanotechnology. As we go forward, we will develop separate module-packets for students and instructors. The instructor information will be more detailed, with additional references and suggestions for integrating the module into existing courses, to ensure that instructors who are not part of the development team have the resources necessary to lead the modules. Student packets will include less detailed write-ups but additional links to videos, background information, and reference materials. Both institutions are committed to continuing to offer these modules and courses. We will follow up with students to assess long-term impacts. We are also initiating an ancillary project developing materials to assist technically trained faculty members, who may not have a strong background in the formal study of ethics, to master the material and infuse active-learning modules in their courses.

Assessment information: The modules were evaluated according to how well we met the learning outcomes. There are many components to the evaluation process: evaluation of module design by the academic and industry advisory council, assessments of learning outcomes through in-class assignments, student evaluations of each module when offered, interval evaluations, site visits by an external evaluator, and follow-up evaluations by the academic and industry advisory council. Ongoing assessments during the fall 2013 and fall 2014 courses at Texas State were largely positive, and assessments of the summer 2013 online course were quite positive. These assessments, both interval and end-of-term, focused on student understanding, engagement, and satisfaction (the latter two are strongly correlated with positive learning outcomes). At UT-Tyler, 87–93% of respondents rated the course good or excellent on a 5-point scale; there were no ratings of fair or poor. At Texas State, evaluations ranged from a high in which 93% of students rated the modules good or excellent and none rated them fair or poor, to a course in which 11% rated the modules fair or poor. This feedback helped project leaders focus on better integrating instructors not originally part of the project team. In November 2013, six focus groups were conducted at Texas State by Dr. Rita Caso, an expert in program assessment and academic evaluation. Participants were students enrolled in courses in which project modules had been presented during the fall 2013 term. On April 23, 2014, Dr. Caso conducted three focus groups with students enrolled in courses at Texas State that incorporated modules: PHIL 1320: Society and Ethics for engineering and engineering technology majors (A-Modules); TECH 4380: Industrial Safety for concrete industry management majors and construction science & management majors; and IE 4380: Industrial Safety for industrial engineering majors (1A and B-Modules). On April 24, 2014, she conducted a focus group with students at UT-Tyler who had completed the B-Module whole course. They reported a high level of interest in nanotechnology and in its ethical implications, and said that outside of class they had encountered issues and information relevant to the material in the modules.

Our assessments show that students are excited about the possibilities of nanotechnology to solve problems and promote better standards of living. Students tell us that the modules have helped them understand the important ethical, sustainability, and social dimensions of emerging technologies, especially nanotechnology. Student feedback is guiding revisions of all modules for future semesters. Student retention is high, and enrollment in the Ingram School of Engineering at Texas State continues to set yearly records.

Additional resources:

Project web page: http://nsf-nue-nanotra.engineering.txstate.edu/home.html

HSI Research Day, Texas State University, San Marcos, March 20, 2013: http://gato-docs.its.txstate.edu/nanotechnology-undergraduate-education/poster.pdf

Micro and Nano Technology Conference: http://gato-docs.its.txstate.edu/nanotechnology-undergraduate-education/Nanotechnology-Safety-Education-Dr-Trybula-As-Presented/Nanotechnology%20Safety%20Education-Dr%20Trybula%20As%20Presented.pdf

Infusing Ethical, Safety, Health, and Environmental Education in Engineering and Technology Curricula, New Horizons in Texas STEM Education Conference: http://nsf-nue-nanotra.engineering.txstate.edu/publications/conferences/contentParagraph/0/content_files/file3/document/Infusing+Ethical%252C+Safety%252C+Health%252C+and+Environmental+Education+in+Engineering+and+Technology+Curricula+-+SA+STEM+Conference.pdf

The Continuing Shock of the New: Some Thoughts on Why Law, Regulation, and Codes Are Not Enough to Guide Emerging Technologies, 121st ASEE Annual Conference and Exposition: http://nsf-nue-nanotra.engineering.txstate.edu/publications/conferences/contentParagraph/0/content_files/file0/document/Continuing+Shock+of+the+New+-+ASEE+presentation.pdf

We Are Seed Planters: A Look at Teaching Students Nanotechnology Environment, Health, and Safety Awareness, Association of Technology, Management, and Applied Engineering: http://nsf-nue-nanotra.engineering.txstate.edu/publications/conferences/contentParagraph/0/content_files/file6/document/2014+ATMAE+CONFERENCE+_Fazarro_.pdf

Enacting Macroethics: Making Social Justice Visible in Engineering Education

Institution: Colorado School of Mines

Faculty/contributors: Jon Leydens, Juan Lucena, Kathryn Johnson

Exemplary features: Focus on macroethics issues in engineering problem defining and solution finding; use of learning progressions to advance ethics knowledge throughout the students' education

Why it's exemplary: Practicing engineers define and solve complex, open-ended, and often ill-structured problems. But undergraduate engineering students get few opportunities in their curriculum to explore open-ended problems or to critically examine the ethical dimensions of engineering problem defining and solving (EPDS) in design. Many problems are predefined and/or closed-ended, so key assumptions embedded in the problem setting are rendered invisible. Those assumptions feature interplays between the technical and nontechnical, particularly of macroethical dimensions of engineering design.

Taking a multicourse, multidisciplinary approach, our program focuses undergraduate students' attention on the complexity inherent in problem setting so macroethical assumptions common to actual EPDS become visible. Working against disciplinary silos, our approach emphasizes (a) macroethical issues of social justice, (b) macroethical assumptions in EPDS processes, and (c) interplays between technical and nontechnical dimensions of EPDS in design.

Program description: Students and faculty work collaboratively so that the undergraduate engineering education experience fosters rich, enhanced, integrated engineering science, design, and humanities/social science experiences. Although roughly 30–40 students are enrolled in the Humanitarian Engineering program minor in any given year, many more students benefit from the enacting macroethics initiative by taking the courses described below as electives. Each year, over 350 undergraduates and 10–12 instructors participate in the enacting macroethics initiative (shown in table 1). Participants also include communities and marginalized groups that engineers sometimes neglect to serve as well as corporate clients with whom students interface.

When students complete our program, they should be able to identify macroethical, social justice issues that are inherent in assumptions made in EPDS processes. Herkert (2005, p. 373) has clarified the distinction between micro- and macroethics: "'Microethics' considers individuals and internal relations of the engineering profession; 'macroethics' applies to the collective social responsibility of the profession and to societal decisions about technology." Although microethics is not ignored, macroethics remains a primary focus of our initiative. Students also show evidence of recognizing and reflecting on the interplays between technical and nontechnical dimensions of EPDS across diverse cultural, ethical, and interdisciplinary contexts. In other words, these soon-to-be engineers practice sociotechnical EPDS.

TABLE 1 Participants in the Enacting Macroethics initiative per academic year

Course	Number of Students	Number of Faculty
EENG 307	55	1
EGGN 301	25	1
EGGN 401	25	1
EGGN 492	150	3–5*
LAIS 377	50	2
LAIS 425	25	1
LAIS 478	25	1
TOTAL	355	10–12

* *Faculty social context consultants*

Our approach is distinctively interdisciplinary and cross-curricular, with one or more courses in the engineering sciences, design, and humanities/social sciences. One course is in the engineering science core: students in mechanical and electrical engineering are required to take EENG 307: Introduction to Feedback Control Systems (IFCS), a third-year course with a section that uses two recurring examples of control systems—in wind energy and active prosthetics—to convey both the technical course concepts and the degree to which social justice dimensions are inherent in defining and solving control systems problems. Although IFCS is not currently required for students in our HE program, the courses mentioned below are all either required or on a menu of options.

A design sequence exemplifies another approach to macroethical instruction. Students begin by taking EGGN 301: Human-Centered Problem Definition, where they learn to place users' perspectives at the center of defining problems by developing listening and empathy skills in order to define problems with (not for) others. With that foundation, they take EGGN 401: Projects for People, where they further define design alternatives, paying close attention to what key stakeholders want and care about and to what will contribute to both their and societal well-being. Finally, in EGGN 492: Senior Design, students work on one of the HE projects such as designing bikes for persons with disabilities, energy efficiency of Native American houses, and prenatal technologies for low-income mothers. Senior Design teams also work with social context consultants, who use a Socratic approach by raising some of the most relevant macroethical, social justice–related questions described below.

Courses in the humanities and social sciences place these macroethical questions in the context of actual case studies. For instance, in LAIS 478: Engineering and Social Justice, students learn to identify and challenge the engineering mindsets and ideologies that get in the way of engineers becoming agents for social justice. They also question how these mindsets contribute to the exclusion of macroethical concerns in problem definition and solution. In LAIS 425: Intercultural Communication, students learn to identify nuanced assumptions embedded in EPDS as they emerge from national, ethnic, ethical, and other normative frameworks. In LAIS 377: Engineering and Sustainable Community Development, students learn to move beyond the limitations of existing engineering problem-solving methods and apply criteria for sustainable community development to engineering projects in order to assess how they contribute to communities' well-being.

In the IFCS design course sequence and in the HSS courses, we aspire to have students explore as many of the following Enacting Macroethics initiative questions as possible:

- In talking with your clients or community partners, what forms of listening enabled you to understand their needs, desires, and aspirations? How did this listening impact your process of defining and later solving the problem?
- What social structural conditions maintain conditions of inequality, and how might your design address such conditions?
- How have you understood a community's political agency and the resources the community members can leverage to carry out, develop ownership of, and maintain the project over the long term?
- What resources and opportunities has your design helped create or could it help create?
- What risks and harms—technical, social, cultural, ethical—has your design intentionally sought to preclude?
- And most importantly, what human capacities has your design endeavored to enhance?

These questions act as heuristics to guide the analysis of engineering case studies and of student EPDS design activities. The final question builds primarily from the work of Nussbaum, which provides a clear end goal for macroethical work.

Assessment information: Quantitative and qualitative educational research methods have facilitated student learning assessment across multiple curricular spaces. For instance, in EENG 307: IFCS, student surveys helped establish a baseline on students' prior exposure to macroethical, social justice issues and their preclass understanding of the meaning of social justice. Quantitative analyses have shown that a majority (71%) of respondents report having been exposed to social justice in their courses at CSM. Also, more than 80% of respondents considered it somewhat or very appropriate for professors to teach social justice concepts in both technical and nontechnical classes and for practicing engineers to consider social justice when designing engineering solutions. Qualitative research using grounded theory methods includes semistructured focus groups and interviews. Findings of the initial qualitative analysis (fall 2014) indicate that some students report a need to switch mental gears when moving between technical and social factors in engineering; that the professor's attempts to connect course material to real-world applications may be too abstract for some students; and that many students appreciated the efforts to integrate social justice into the course, partly because they felt it would provide leverage for learning technical elements. The fall 2015 IFCS iteration aims to directly address these issues; using wind energy and active prosthetics as recurring examples across the course, we are assessing degrees of improved learning of multiple technical and macroethical course concepts.

Across courses in the Enacting Macroethics initiative, evidence of student learning includes cognitive and attitudinal dimensions. Several courses include pre- and postcourse evaluations measuring student understanding of key complex concepts and interrelations (e.g., between engineering and social justice, their willingness to engage social justice through engineering practice, and how after courses they see their career alternatives in a different light). Each course has final projects and/or presentations that act as summative assessment mechanisms. For instance, in EEGN 307: Introduction to Feedback Control Systems, final projects involve an investigation of a real-world control system and its broader social justice implications. In LAIS 425: Intercultural Communication, students complete pre- and postcourse video self-interviews using the same question prompts; in a final paper, students identify the differences between the two self-interviews, particularly key cognitive and attitudinal shifts.

Beyond course-level assessment, evidence of the Enacting Macroethics initiative's impact also emerges through institutional support for the HE program in which it is housed. Evidence suggests shifts in our institutional culture, in a university with deep connections to extractive industries and fossil fuels. Thanks to the HE program, we now enjoy a regular and well-funded lecture series (4–6 lectures per year) that engages corporate, NGO, and academic actors in analyzing, for example, the social justice dimensions of mining on nearby communities. Furthermore, career services, fund raising initiatives, and recruitment/retention programs have begun to focus on the HE program as an instrument for progressive institutional change. This has resulted in more than $500K in gifts from donors who realize the potential of the HE program for the ethical education of engineering graduates.

Additional resources:

Johnson K, Leydens JA, Moskal BM, Silva D, Fantasky JS. 2015. Social Justice in Control Systems Engineering. Presentation at the ASEE Conference, June 14–17, Seattle. Available at https://peer.asee.org/social-justice-in-control-systems-engineering

Leydens JA, Lucena JC. 2014. Social justice: A missing, unelaborated dimension in humanitarian engineering and learning through service. International Journal for Service Learning in Engineering 9(2):1–28. Available at http://library.queensu.ca/ojs/index.php/ijsle/article/view/5447

Leydens JA, Lucena JC, Nieusma D. 2014. What is design for social justice? Presentation at the ASEE Conference, June 15–18, Indianapolis. Available at https://peer.asee.org/what-is-design-for-social-justice

Lucena JC, Leydens JA. 2015. From sacred cow to dairy cow: Challenges and opportunities in integrating of social justice in engineering science courses. Presentation at the ASEE Conference, June 14–17, Seattle. Available at https://peer.asee.org/from-sacred-cow-to-dairy-cow-challenges-and-opportunities-in-integrating-of-social-justice-in-engineering-science-courses

Lucena JC, Schneider J, Leydens JA. 2010. Engineering and sustainable community development. Synthesis Lectures on Engineers, Technology, and Society 5(1):1–230. Available at www.morganclaypool.com/doi/abs/10.2200/S00247ED1V01Y201001ETS011

Ethics When Biocomplexity Meets Human Complexity (Role-Play Workshop) and Nanosilver Linings Case

Institution: Indiana School of Medicine-South Bend and University of Notre Dame

Faculty/contributors: Kathleen Eggleson, Indiana School of Medicine-South Bend (formerly at University of Notre Dame); Joshua Dempsey (student), University of Notre Dame

Exemplary features: Interactive and creative education approach; consideration of macroethics issues

Why it's exemplary: The Nanosilver Linings case and the workshop, "Ethics When Biocomplexity Meets Human Complexity," that supports it are exemplary because they are based on best practices in the field (e.g., clear definition of learning objectives, active learning, interactive learning, case-based learning, role play), provide instructors with refined and assessed (by both student participants and an external faculty expert) materials sufficient for a 3-hour ethics education workshop, and offer students the experience of STEM-relevant role play with richly detailed stakeholder characters in a realistic hypothetical case. The robust supporting materials provide an organized reading list, instructor checklist, time table, slides, and guidelines for role play. This activity is sponsored by the National Science Foundation (Award #1338682), for the Ethics Education in Science and Engineering (EESE) Program. It is a product of the Collaborative Research project Ethics Education in Life Cycle Design, Engineering, and Management.

Program description: Participants are graduate students in any field of science or engineering. The case and workshop have been piloted and refined through initial offerings to cohorts of STEM graduate students at two universities. In addition, the Nanosilver Linings case has been offered at an academic research institute for both faculty and students, spanning STEM disciplines and STEM-related fields (e.g., science policy), and went smoothly and was well received by participants. Learning objectives underpinned the design of the case and workshop, and their achievement was assessed formally through the instrument administered upon completion of the workshop. Through this workshop, STEM graduate students learn to:

- List ethical dilemmas involved in public communications about science and technology
- Appreciate the human factors, conflicts of interest, struggles, and tradeoffs in a participatory governance scenario pertaining to science and technology
- Identify stakeholders in complex decisions pertaining to science and technology
- Understand how the perspectives of different stakeholders are informed and communicated
- Understand the inherent limits of quantitative, technical methods of assessment in incorporating values
- Operate professionally as a scientist or engineer even in "grey areas" of practice where there is no possibility of a single correct answer

These learning objectives prepare natural and applied scientists for ethical research, practice, and leadership. For example, on the assessment instrument, in response to the question "What event during the workshop changed your thinking? In what way did your thinking change?" one student answered "Discussion of our responsibility as scientists to be ambassadors to the general public. I have a responsibility. I need to do my due diligence as an academic."

Methods and content: The Nanosilver Linings role play case, delivered through the workshop, provides science and engineering graduate students with an active learning experience on the "wicked problems" of emerging technology macroethics. Participants play one of seven societal stakeholders in a hypothetical scenario involving the possible location of a nanosilver food packaging company in an economically struggling city. Both social and scientific implications are considered around the product life cycle, during the role play and in structured discussion when participants are out of character. The event calls on participants to practice intellectual integration of technical, moral, legal, and societal aspects of a complex science/technology situation as well as spontaneous interpersonal communication—skills that will be useful in myriad aspects of their careers.

To further elucidate methodology, an excerpt of the Instructor Notes for Workshop Leader is included here:

This is a role play workshop designed for ethics education of STEM graduate students. It primarily emphasizes societal-level macroethics related to decision making related to commercial application of emerging nanotechnoloies, as opposed to microethics or responsible conduct of research (RCR). However, students will confront dilemmas at the level of individual contact through perspective-taking in acting as one of seven characters in a hypothetical, but realistic, case. To offer the Nanosilver Linings case in the context of one, three-hour workshop, the basic steps are:

- Register 7 students per group. (The workshop can run with either 6 or 7 students, allowing room for one cancellation or no-show without disrupting the role play case.) Doodle internet polling can be used for this purpose, choosing the (free) option to limit the number of participants.
- Prepare materials (copies of the Nanosilver Linings case, character folders including readings and private information, nametags, certificates, assessment forms).
- One week in advance, send out the set of readings intended for all participants.
- Adapt workshop slides with photos of your registered participants.
- Water/coffee and baked goods may be served during the event.

Requirements

- Groups of 6 or 7 participants are required for this exercise. It is recommended that, for a free-standing workshop, 7 participants be scheduled in advance; that way if there is a cancellation or no-show on the day of the event, the workshop can take place without need for recruiting a substitute on short notice.

Options and Flexibility Personnel

- The character Carlson, concerned parent, may be included or excluded, allowing a ±1 extent of flexibility in number of participants per group.
- Participants may be engaged in the study of any STEM or STEM-related field (e.g., philosophy of science, science policy).
- Participants may be from the same or different fields.
- Participants may be at different levels of study; this experience was designed with STEM graduate students at any level or year of study in mind, but may also be appropriate for advanced undergraduates.
- Participants may know one another well, or not at all, prior to the workshop.
- Characters' assignments may be determined by random draw, by the workshop leader, or by the participants.

Time

- Running time may be adjusted through time allotted for reading, accordingly adjusting the amount and difficulty of readings selected or assigning readings in advance.
- Electronic highlighting can be applied to readings before printout to draw out the most pertinent passages, thus reducing reading time and volume while maintaining the original document context.
- Time allotted for discussion is flexible, and can be used to adjust total running time.
- The length and nature of the break is flexible.

Content

- Selection of readings by the workshop leader allows flexibility with regard to (a) level of difficulty and (b) subject matter emphasis.

Materials Checklist

- Informed consent form, if applicable
- Identical initial packets for each participant, with case plus selected readings
- Slides with character identities and student photos (prepared while participants are in common learning phase); template provided in Power Point file
- Character nametags
- Character-specific packets, with character information and selected readings
- Discussion questions/slides (Power Point file)
- Assessment forms

Assessment information: (1) Quantitative and (2) written responses on assessment instrument, (3) external evaluator Michael Loui (formative and summative involvement), and (4) focus group. (1) On a 5-point Likert scale, where 5 is strongly agree and 4 is agree, graduate student participants across four cohorts (n=26) agreed with the following statements: I would recommend this experience to other STEM graduate students (4.69), This experience makes me more aware of my own values as they pertain to science and engineering applications (4.62), This experience was a good use of my time (4.58), and This experience makes me more aware of the values of other people as they pertain to science and engineering applications (4.5). Where 5 is highly satisfied and 4 is satisfied, students were satisfied with the realism of the hypothetical case (4.69) and the appropriateness of readings for character (4.42). (2) In answer to the question: What was the most surprising thing you learned from the workshop?, one student said "Most of the characters had a bias/motivation to be biased to benefit themselves in the situation. I think this highlights the need for ethical, unbiased work to represent truth/underrepresented populations." Some of the insights shared in response to this question were fundamental: "Grey things can be 'made' completely black or completely white depending on how you want to use the information"; "Making decisions in the 'real world' is not as black and white as I had initially thought. Much more goes into everyone's decisions." For the question, What event during the workshop changed your thinking? In what way did your thinking change?, one student replied "When we were speaking about the responsibilities of the small community to make decisions that impacted the future of the community/larger scope society with limited representation. It is hard to understand/think about this, since in a way, it makes us all responsible for each other, even though we don't act like it." Other responses to this question included: "Thinking about stakeholders not represented in the workshop then discussing who they were/possible pros and cons that could impact them. Usually this isn't discussed, and thinking about it is important!" and "Discussion of our responsibility as scientists to be ambassadors to the general public. I have a responsibility. I need to do my due diligence as an academic." (3) Excerpts from the external evaluator's report: "The positive comments from the focus group indicate that the current version of the workshop is engaging and appropriately challenging." "Overall, I believe you have designed an intellectually challenging, emotionally engaging, and likely enjoyable experience that teaches students to consider the variety of stakeholder viewpoints in making ethically difficult decisions about technology and society." (4) Feedback from focus group participants, as reported by the external evaluator: "Students strongly agreed that this workshop format was far superior to the one-day all-campus RCR training because the content was more useful, practical, and directly relevant to science and engineering, and because the workshop required active participation: it required more thinking about the challenging ethical issues."

Additional resources:

Ethics when Biocomplexity meets Human Complexity Role Play Workshop and Nanosilver Linings Case: https://nationalethicscenter.org/resources/7811

Creating a Community of Ethics Educators in Engineering

Institution: Pennsylvania State University

Faculty/contributors: Thomas Litzinger, Nancy Tuana, Xiaofeng Tang

Exemplary features: Approach to preparing engineering faculty to teach ethics that is integrated and relevant to engineering

PennState

Why it's exemplary: For more than a decade the Leonhard Center and Rock Ethics Institute have collaborated to design and implement activities aimed at incentivizing, preparing, and supporting engineering faculty to integrate ethics in their teaching. These activities are exemplary in three aspects: (1) Interdisciplinary—Led by a team of philosophers and engineers, the initiatives combine skills such as ethics spotting and ethical decision making with relevant factors of teaching, advising, and research in engineering. (2) User-oriented—Instead of presenting prepackaged teaching in a top-down approach, the initiatives are firmly grounded on the actual needs and challenges perceived by engineering faculty (e.g., big lecture courses vs. small project-based courses) and provide them with the skills to integrate ethics in their teaching. (3) Across the curriculum—The initiatives serve the specific and relevant needs of engineering instructors who teach a variety of courses, from first-year seminars to senior capstone design and graduate courses.

Program description: Since 2002 the Leonhard Center for Enhancement of Engineering Education and the Rock Ethics Institute at Penn State University have worked together to engage and support engineering faculty members in integrating ethics in their teaching. Drawing on their respective strengths in engineering education and ethics education, the two institutes have launched a series of faculty development initiatives aimed at creating a community of ethics educators in engineering. Over more than a decade these initiatives have provided basic training in ethics skills and instructional design to more than 100 participants at and beyond the College of Engineering at Penn State, serving every discipline of engineering. Three particular initiatives symbolize the two institutes' commitment to engaging engineering faculty in ethics education: Learning and Teaching Ethics in Engineering, Creating an Ethical Classroom, and Enhancing Ethics Education of Graduate Students. From 2002 to 2010 the Leonhard Center and the Rock Institute offered eight summer workshops on "Learning and Teaching Ethics in Engineering," designed with three initial objectives: (1) preparing the participants (engineering faculty members) to teach ethics in engineering; (2) helping participants design ethics-related course activities specifically tied to the content in major courses with instructional design methods; and (3) promoting ethics education throughout the college. The workshops were designed and facilitated by the directors of the two institutions as well as invited faculty members from the Philosophy Department and College of Engineering. The workshops presented in accessible language the basic concepts and frameworks of ethics because many engineering faculty members expressed concern and anxiety about their lack of understanding of ethics. Leaders of the workshop also shared their teaching experiences and helped the faculty participants apply instructional design methods to create course assignments and discussions for integrating ethics. The 3-day workshops were supplemented with two follow-up meetings: one month after each workshop, the participants met to present the learning objectives, instructional strategies, and assessment methods each had chosen for their own courses and received feedback from peers and from workshop leaders; several times there was an additional meeting at the end of the spring semester the following academic year, where all the participants met again to present the implementation of the ethics teaching plan in their own courses and evaluate their experiences. Since 2013 the Leonhard Center has collaborated with Dr. Tricia Bertram Gallant, an internationally known expert on academic integrity, to offer workshops entitled "Creating the Ethical Classroom" for faculty at the College of Engineering. The workshops have three objectives: (1) to enhance participants' self-efficacy in teaching academic integrity and professional ethics; (2) to help participants develop instructional strategies to integrate academic integrity and ethics in their courses; and (3) to support faculty in implementing the ethics-related teaching plan in classes. Drawing on Dr. Bertram Gallant's long experience in engaging issues involving academic integrity and helping students understand its importance, the workshops help the faculty participants understand the ways in which choices that they make in their assignments and assessments affect students' behavior. The workshop also discusses the literature—and students' perspectives—on academic integrity. The workshops provide strategies for instructors to introduce academic integrity not as arbitrary rules but as principles closely related to the development of professional engineers. For example, the participants receive suggestions on helping students understand the values underlying academic integrity and how many of the same values underlie engineering codes of ethics. In the month after the workshops, faculty participants develop their own plans to integrate academic integrity more effectively in their teaching and to create bridges to professional ethics. Before the fall semester starts, the Leonhard Center hosts meetings for faculty participants to share their plans and receive feedback from the Center director, peer workshop participants, and Dr. Bertram Gallant via video chat. At the end of the following academic year, faculty participants meet again to report and evaluate their implementation of the teaching plan.

In fall 2014 the Leonhard Center and Rock Ethics Institute began a new initiative focused on ethics education for graduate students, jointly employing a postdoctoral scholar in engineering ethics to expand ethics education in the College of Engineering, beginning with an assessment of the challenges and needs of ethics education at the graduate level. In spring 2015 the postdoc interviewed graduate coordinators and faculty representatives from all but one graduate program in the College of Engineering. The interviews explored current approaches to ethics education in each graduate engineering program and examined the advantages and challenges of the approaches and needs of different programs. As Penn State University requires all of its graduate students to complete Scholarship and Research Integrity (SARI) training, the interviews also inquired into the status of this training in each program. Findings of the current approaches, challenges, and needs of graduate ethics education have been summarized and reported to the associate dean of the college. The Leonhard Center and the Rock Institute are scheduling a meeting with all the engineering graduate coordinators to present these findings and to propose collaborative projects to build resources (e.g., online learning modules) to assist graduate ethics education in engineering. An initial activity of the team was to create a 4-hour ethics workshop for graduate students involved in an NSF ERC on medical devices. In this workshop, the students were asked to create visual representations ("connections maps") of the many connections of their particular research—e.g., developing a new biosensor—to other researchers, users, and patients that might one day use or be affected by their devices. Students also considered aspects of the production of medical devices and impacts on people and the environment. These diagrams were then used to explore different ethical issues involved in human-human and human-environment interactions.

Assessment information: Over the years of our work on ethics, our assessment has improved in sophistication. Early on, we used surveys of participants and asked about the effectiveness of the workshops and whether they were meeting the needs of the participating faculty. Survey results provided a basis for improving the workshop design. After offering the "Learning and Teaching Ethics" workshop for several years, we interviewed past participants to learn what they found most useful and what they were still using from the workshops. Most were still using what they had learned in the workshops; the specific tools they were using depended on the way they had decided to integrate ethics in their courses. For example, some participants had their students study the introduction to ethical frameworks from the workshop and use the frameworks to analyze codes of ethics. Others used models for ethical analysis of cases that were presented and used in the workshop.

Our assessment of "Creating an Ethical Classroom" involves pre- and postworkshop surveys and interviews of the participants as well as surveys and short-answer assessments of their students. Data from these assessments show that the workshops are seen as very valuable by the participants, changing their perspectives on academic integrity from compliance to a perspective of trying to inspire students to act with integrity. Conversations with faculty about the workshops indicate that beginning with academic integrity and bridging to professional ethics is an intellectually comfortable way for them to engage ethics. Postworkshop interviews showed that participants felt more confident discussing academic integrity topics with students after the workshop; participants also reported instructional changes to integrate academic integrity in a variety of ways—syllabi, class discussion, course assignments, and exams.

To assess the outcomes of the faculty development on students' ethical learning, pre- and postsurveys were conducted in faculty participants' classes. Published research outcomes show that students developed deepened understanding of academic integrity and its importance for engineering professional development. Students also acknowledged the effectiveness of class discussion of academic integrity and clearly perceived the instructional changes compared with other courses.

Since 2013 the assessment for the "Creating the Ethical Classroom" workshop has evolved. For the summer 2015 workshop faculty participants are interviewed three times: before the workshop, immediately afterward, and one after the faculty members have implemented the ethics teaching in their courses. Feedback generated from the more comprehensive assessment will be used to continuously improve the faculty development initiative. The new initiative for expanding ethics education at the graduate level is still at an early stage, but the interview data for the current ethics education, strengths, and limitations in the engineering graduate programs will serve as a "benchmark" for comparison once new ethics programs are implemented.

Ethics Sessions in a Summer Undergraduate Research Program

Institution: University of Illinois at Urbana-Champaign (UIUC)

Faculty/contributors: Michael C. Loui, UIUC and Purdue University

Exemplary features: Infusing ethics into a NSF Research Experiences for Undergraduates (REU) Sites program and critical assessment that reveals areas for improvement

Why it's exemplary: Our ethics program consists of six interactive 1-hour sessions in which small groups of 3 to 5 students discuss short cases that are fictional but realistic. The case topics are selected to be relevant to the students' interests. Taught with a general approach to ethical reasoning that uses everyday language rather than abstract philosophical principles, students gain skill in ethical reasoning through repeated practice, with active learning in small collaborative groups. They are assessed through pre- and post-tests using a counterbalanced design. Each test requires the analysis of a case that is scored with a common rubric that aligns with the learning objectives. This ethics program can be easily integrated into a summer undergraduate research program. The program's small-group pedagogy can be scaled up to student groups of any size in any instructional setting with minimal changes.

Program description: In the summers of 2009–2012 the Information Trust Institute at the University of Illinois at Urbana-Champaign hosted an 8- to 10-week summer undergraduate research program on reliable and secure computing, supported by a grant from the Research Experiences for Undergraduates (REU) Sites program of the National Science Foundation (grant CNS-0851957). Most of the 21–26 students were majoring in computer science, computer engineering, electrical engineering, or another technical discipline. Each summer included 6 weekly sessions on ethics in the responsible conduct of research (RCR) and in the development and use of computing technology.

The sessions addressed both micro- and macroethical topics such as professional responsibility, authorship, plagiarism, mentoring relationships, conflict of interest, software quality, privacy of personal data, confidentiality of intellectual property, accuracy of computational models, and social impacts of computers. We chose these topics for their relevance to the students' research projects. We omitted standard RCR topics that were not relevant to these students, such as the responsibilities of peer reviewers and the protection of human and animal subjects. Even the traditional RCR topics of fabrication, falsification, and data management were not relevant for many projects that involved the development of software or the mathematical analysis of algorithms. We selected fictional but realistic short cases (scenarios) from a variety of sources, including textbooks on computer ethics and the NAE's Online Ethics Center for Engineering and Science. In 2011 and 2012, we replaced the session on ethics in computational modeling by a showing and discussion of the 36-minute movie "Henry's Daughters," which highlights ethical issues in a dramatized case in which engineers design an intelligent transportation system with autonomous vehicles. In ethics presentations for other REU site programs in the summers of 2013 and 2014, after the Information Trust Institute's REU grant had ended, we replaced some of the RCR cases with short videos (less than 4 minutes) developed at the University of Nebraska-Lincoln. We substituted the video cases for text cases because we expected that students would find video cases more interesting and memorable. Our expectations were confirmed in the program evaluation surveys at the end of each summer (not reported here).

The ethics sessions used active learning methods: collaborative and cooperative learning. We chose active learning through small-group discussion because, as Wilbert McKeachie and Marilla Svinicki have written in their book *Teaching Tips*, "Discussion methods are superior to lectures in student retention of information after the end of a course; in transfer of knowledge to new situations; in development of problem solving, thinking, or attitude change; and in motivation for further learning." In each 60-minute ethics session the students were randomly divided into small groups of 3–5 students to simultaneously read and discuss the same case for about 10 minutes. Then a professor led a discussion of this case with the entire cohort. During this discussion period, he invited different groups to respond to questions about the case for about 10 minutes. The students were asked to identify the ethical issues and to suggest what the characters in the case should do next, for what reasons. Then the session moved on to another case, again with simultaneous discussions in small groups followed by a discussion with the entire cohort. One session was organized differently: Each small group took responsibility for reading and answering questions about one of five cases dealing with the social impacts of computers. For the first 10 minutes, all five groups read and discussed their case simultaneously, then the professor interacted with each group in turn to discuss that case while the other groups listened.

At the beginning of the first ethics session of the summer program, we presented a general approach to ethical problems. Our general approach uses everyday language because, with limited time in a summer REU program, students need guidance in thinking about ethics issues without having to learn philosophical jargon.

A General Approach to Ethical Problems (1) Identify the affected parties, their interests (rights, expectations, desires), and their responsibilities. Determine what additional information is needed. (2) Consider alternative actions by the main actors, and imagine possible consequences. (3) Evaluate actions and consequences according to basic ethical values—honesty, fairness, trust, civility, respect, kindness, etc.—or the following tests: (a) Harm test: Do the benefits outweigh the harms, short term and long term? (b) Reversibility test: Would this choice still look good if I traded places? (c) Common practice test: What if everyone behaved in this way? (d) Legality test: Would this choice violate a law or a policy of my employer? (e) Colleague test: What would professional colleagues say? (f) Wise relative test: What would my wise old aunt or uncle do? (g) Mirror test: Would I feel proud of myself when I look into the mirror? (h) Publicity test: How would this choice look on the front page of a newspaper? Each student received the Association for Computing Machinery code of ethics, a book chapter on ethics for computing professionals by Deborah Johnson and Keith Miller, and a copy of the third edition of the booklet *On Being a Scientist*, an overview of RCR by the National Academies of Sciences, Engineering, and Medicine. Students were not tested on these readings, however, and they were not assigned any other ethics homework. As learning objectives, through the ethics sessions, we expected students to learn to identify the ethical problems or dilemmas, recognize the people affected and understand their perspectives, identify a comprehensive list of actions, and provide a justified action to resolve the ethical problem or dilemma.

Assessment information: To assess the effectiveness of the ethics sessions, we asked students to analyze two short cases. Case A highlighted ethical issues in computing technology, and case B raised ethical issues in conducting research. The students were randomly assigned to two groups in a counterbalanced pre-/post-test design. One group received case A for the initial assessment at the beginning of the summer and case B for the final assessment at the end of the summer; the other received case B initially and case A at the end. For each case, students responded to four questions, which corresponded to the four intended learning objectives: (1) What ethical issues does this case raise? (2) Who is affected by this case? What are their perspectives on the case? (3) What actions might the characters consider to resolve the ethical issues? (4) Among these actions, which should the characters choose? For what reasons? These questions followed our general approach described above. For each assessment, students were expected to take 30–60 minutes, working individually and without consulting any references. There was no limit on the lengths of their responses, which were independently scored by two evaluators using a common rubric that specified three performance levels for each of the four questions. They compared their scores and discussed any differences. After discussion and reconciliation, the scores differed by at most one point on each question. The scores were combined to obtain a cumulative score for each student. In the summer of 2009, we had initial and final responses for 17 students. In the summer of 2010 we had initial and final responses for eight students. Because the numbers of students were small, we aggregated the 2009 and 2010 data by case. We used the Mann-Whitney *U* test for independent samples to analyze the differences between the initial and final responses because the data did not pass the Shapiro-Wilk normality test or a test of homoscedasticity. We found no significant differences between the initial and final scores for case A or for case B. We suspect that there was essentially no difference in the initial and final scores because the content of the ethics sessions was not formally reinforced outside of the sessions through additional academic work. In addition, the ethics sessions might not have added significantly to the knowledge and skills of the students who had previously taken computer ethics courses that were required in their undergraduate computer science programs. At the end of the summer, the students probably put minimal effort into the post-test. Finally, our intended learning outcomes may have been too ambitious, and thus the assessment task was too difficult. As a consequence, students might have been unable to demonstrate what they had learned. We believe that our assessment method can be applied broadly. As our experience suggests, however, even when the ethics sessions are taught with appropriate pedagogies, and when the assessments are aligned with the learning objectives, students might not demonstrate improved skills in analyzing ethics cases.

Additional resources:

Cooperative Learning and Assessment of Ethics Sessions in a Summer Undergraduate Research Program: https://uofi.box.com/s/h350iv2mn2p4m9o0blbxls0o6hzjcw6h

Appendix A: Thank You to All the Contributors

The Selection Committee acknowledges the work of the following individuals and their activities.

A Course on "Social Media and Public Health" at SUNY-Albany: Ricky Leung, SUNY-Albany
A Graduate Course Using a Conceptual Model to Identify the Linkages among Technology, Economics, and Societal Values: Otto Loewer, University of Arkansas
Anticipatory Engineering Ethics: Richard Wilson, University of Maryland–Baltimore County
Case Studies for Engineering Ethics across the Product Life Cycle: Matthew Eckelman, Northeastern University
Corporate Social Responsibility Course: Jessica Smith, Colorado School of Mines
Creating a Community of Ethics Educators in Engineering: Thomas Litzinger, Penn State University
Cyber Aggression and Cyber Warfare: An Anticipatory Ethical Approach: Richard Wilson, University of Maryland–Baltimore County
Enacting Macroethics: Making Social Justice Visible in Engineering Education: Jon Leydens, Colorado School of Mines
Engineering a Catastrophe: Ethics for First-Year STEM: Tobias Rossmann, Lafayette College/Rutgers University
Engineering Ethics in Context: Brent Jesiek, Purdue University
Ethical Autobiography: Sandra Woodson, Colorado School of Mines
Ethical Issues in Software Design Course: Chuck Huff, St. Olaf College
Ethics Activities in the Civil Engineering Curriculum at the United States Coast Guard Academy: Hudson Jackson, United States Coast Guard Academy
Ethics and Engineering for Safety: Nancy Leveson, Massachusetts Institute of Technology
Ethics as Philosophical History for Engineers: Daniel Biezad, California Polytechnic State University
Ethics Sessions in a Summer Undergraduate Research Program: Michael Loui, University of Illinois at Urbana-Champaign and Purdue University
Ethics When Biocomplexity Meets Human Complexity (Role-Play Workshop) and Nanosilver Linings Case: Kathleen Eggleson, Indiana School of Medicine–South Bend and University of Notre Dame
Foundations of Global Leadership Course: Gregg Warnick, Brigham Young University
Global Engineers' Education Course: Bhavna Hariharan, Stanford University
Global Trends: Strategic Analysis and Systems Thinking for Leadership: Darryl Farber, Penn State University
Graduate Course in Research Ethics: Carl Mitcham, Colorado School of Mines
Humanitarian Engineering, Past and Present: A Role-Playing First-Year Course: Kristin Boudreau, Worcester Polytechnic Institute
Introduction to Software Engineering Ethics: Irina Raicu, Santa Clara University
Learning to Listen: A Tool for Morally Engaged Engineering Practice: Yanna Lambrinidou, Virginia Tech
Line Drawing Technique: Ashraf Ghaly, Union College
Mock Internship Hiring Activity: Matthew Jensen, Florida Institute of Technology
Multiyear Engineering Ethics Case Study Approach: Daniel Saulnier, Northeastern University
NanoTRA: Texas Regional Alliance to Foster Nanotechnology Environment, Health, and Safety Awareness in Tomorrow's Engineering and Technology Leaders: Craig Hanks, Texas State University
Nature and Human Values Course: Sarah Jayne Hitt, Colorado School of Mines
Partnership for Global Health Technologies: Katie Clifford, Boston University
Phenomenological Approach to Engineering Ethics Pedagogy: Valorie Troesch, Michigan Technological University
PRIME Ethics: Purdue's Reflective & Interactive Modules for Engineering Ethics: Andrew Brightman, Purdue University
Problem-Based Learning in a Professional Ethics Course for Undergraduate Engineering Students: Robert Kirkman, Georgia Institute of Technology
Professional Aspects of Engineering (Graduate Course EGR 602): Nael Barakat, Grand Valley State University
Responsibility of Engineering: Codes & Professionalism (3-hour university course): Steve Starrett, Kansas State University
Student Ownership of Ethics: Sundararaj Iyengar, Florida International University
Team Ethics Assignment: Based on Engineering Student Co-Op Experience: Laura Grossenbacher, University of Wisconsin–Madison
Terrascope: Ari Epstein, Massachusetts Institute of Technology
The Ethics of Engineering: A Discussion: Barry Belmont, University of Michigan
The Golden Heart Program: Malini Natarajarathinam, Texas A&M University
The University of Virginia SEAS Senior Thesis: A Culminating Activity: Deborah Johnson, University of Virginia
Three Course Sequence in Medical Device Commercialization: Deborah Munro, University of Portland
UnLecture on Software Engineering Ethics: Vignesh Subbian, University of Cincinnati
Using Student-Authored Case Studies to Teach Bioengineering Ethics: Rosa Lynn Pinkus, University of Pittsburgh

Appendix B: Map of Exemplary Programs and Activities

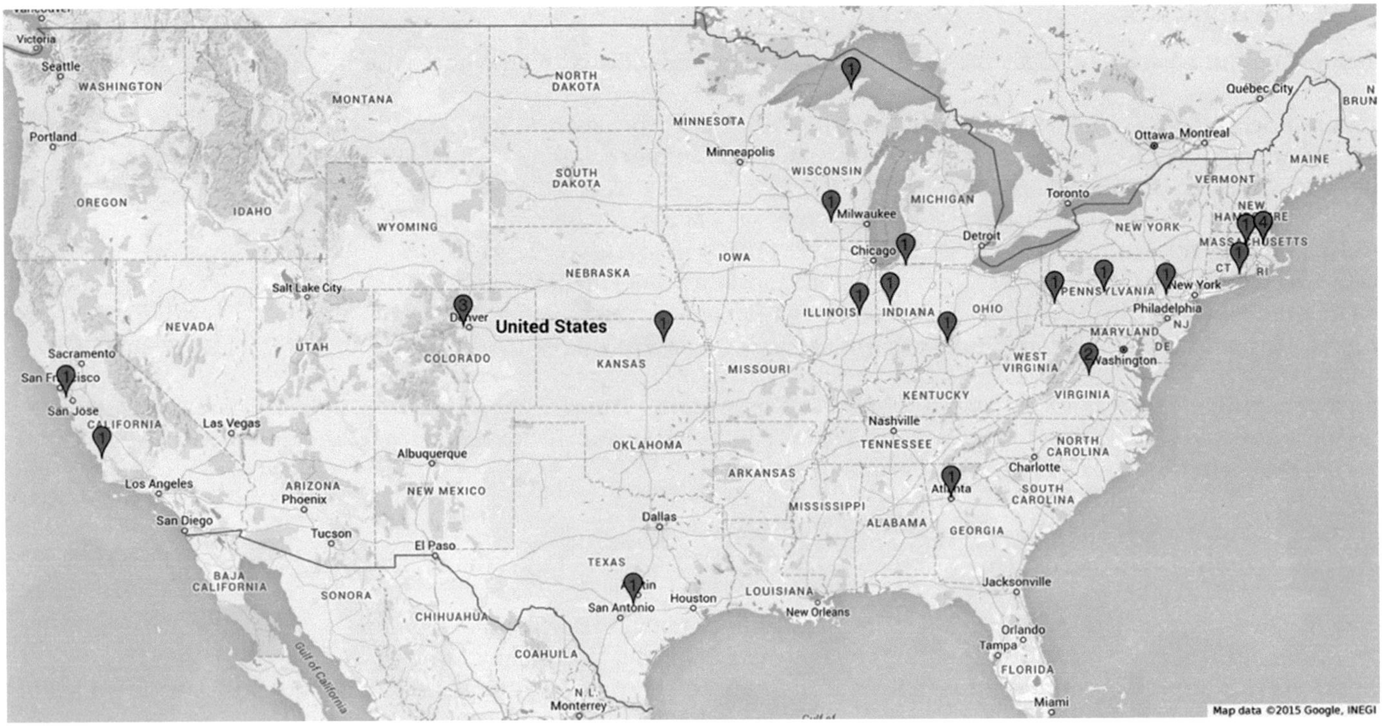

Figure 1 Map of exemplary programs and activities.

Case Studies for Engineering Ethics Across the Product Life Cycle: Matthew Eckeman, Northeastern University
Corporate Social Responsibility Course: Jessica Smith, Colorado School of Mines
Creating a Community of Ethics Educators in Engineering: Thomas Litzinger, Penn State University
Enacting Macroethics: Making Social Justice Visible in Engineering Education: Jon Leydens, Colorado School of Mines
Engineering a Catastrophe: Ethics for First-Year STEM: Tobias Rossmann, Lafayette College/Rutgers University
Ethics Activities in the Civil Engineering Curriculum at the United States Coast Guard Academy: Hudson Jackson, United States Coast Guard Academy
Ethics and Engineering for Safety: Nancy Leveson, Massachusetts Institute of Technology
Ethics as Philosophical History for Engineers: Daniel Biezad California, Polytechnic State University
Ethics Sessions in a Summer Undergraduate Research Program: Michael Loui, University of Illinois at Urbana-Champaign and Purdue University
Ethics When Biocomplexity Meets Human Complexity (Role-Play Workshop) and Nanosilver Linings Case: Kathleen Eggleson, Indiana School of Medicine-South Bend/University of Notre Dame
Global Engineers' Education Course: Bhavna Hariharan, Stanford University
Humanitarian Engineering, Past and Present: A Role-Playing First-Year Course: Kristin Boudreau, Worcester Polytechnic Institute
Learning to Listen: A Tool for Morally Engaged Engineering Practice: Yanna Lambrinidou, Virginia Tech
Multiyear Engineering Ethics Case Study Approach: Daniel Saulnier, Northeastern University
NanoTRA: Texas Regional Alliance to Foster Nanotechnology Environment, Health, and Safety Awareness in Tomorrow's Engineering and Technology Leaders: Craig Hanks, Texas State University
Nature and Human Values Course: Sarah Jayne Hitt, Colorado School of Mines
Phenomenological Approach to Engineering Ethics Pedagogy: Valorie Troesch, Michigan Technological University
PRIME Ethics: Purdue's Reflective & Interactive Modules for Engineering Ethics: Andrew Brightman, Purdue University
Problem-Based Learning in a Professional Ethics Course for Undergraduate Engineering Students: Robert Kirkman, Georgia Tech
Responsibility of Engineering: Codes & Professionalism (3-hour university course): Steve Starrett, Kansas State University
Team Ethics Assignment: Based on Engineering Student Co-Op Experience: Laura Grossenbacher, University of Wisconsin–Madison
Terrascope: Ari Epstein, Massachusetts Institute of Technology
The University of Virginia SEAS Senior Thesis: A Culminating Activity: Deborah Johnson, University of Virginia
UnLecture on Software Engineering Ethics: Vignesh Subbian, University of Cincinnati
Using Student-Authored Case Studies to Teach Bioengineering Ethics: Rosa Lynn Pinkus, University of Pittsburgh